James Temple Brown

The Whale Fishery and Its Appliances

James Temple Brown

The Whale Fishery and Its Appliances

ISBN/EAN: 9783337328924

Printed in Europe, USA, Canada, Australia, Japan

Cover: Foto ©berggeist007 / pixelio.de

More available books at **www.hansebooks.com**

GREAT INTERNATIONAL FISHERIES EXHIBITION.

LONDON, 1883.

UNITED STATES OF AMERICA.

E.

THE WHALE FISHERY

AND

ITS APPLIANCES.

BY

JAMES TEMPLE BROWN,

Assistant in the Department of Art and Industry, U. S. National Museum.

WASHINGTON:
GOVERNMENT PRINTING OFFICE.
1883.

INTRODUCTION.

Pursuant to instructions received from the United States Commissioner of Fish and Fisheries, to take charge of the collection and preparation of an exhaustive exhibit illustrative of the whale fishery, to be displayed at the International Fisheries Exhibition of the present year, I proceeded to the principal whaling ports of the eastern coast. The field work was conducted mainly at New Bedford, Provincetown, Nantucket, Edgartown, and New London. The object of my visits being known, the whalemen, agents, owners of vessels, and others interested in this industry, with one accord, offered their services to the Government, and generously responded to its call, in order that the vast machinery of the whale fishery of the United States might be represented in the friendly contest among all the prominent maritime nations of the world. From the fresh material collected on this tour, as well as from objects previously deposited in the Fisheries Division of the National Museum, selections have been made and prepared for final exhibition in London. In the preparation of the whaling craft it has been my desire that the objects should be exhibited as nearly as possible in the same condition in which they are usually placed on whaling vessels. The only exceptions allowed to this plan of operations occur in regard to several objects nickel-plated by one manufacturer, who is anxious to display his goods in an attractive manner. As is well known, the best kinds of wood, rope, iron, and steel are sought by whalemen, and the manufacturers, either through pride or fear of competition, employ the best grades of material, and finish some of their goods in an artistic manner.

The exhibit embraces, for the most part, the apparatus used at present; but some rare and interesting implements that were hastily constructed on vessels in times of necessity, as well as some that were developed as experiments both at sea and ashore, have also been included. The unique designs of the last-named series afford an interesting study. Some of them are obsolete, while others have developed into more perfect and acceptable forms, and though they have, in part, been superseded by improved contrivances, they have been, nevertheless, actively employed, and are worthy of prominent places.

Several objects made and used by the Eskimo tribes of the Hudson Bay region appear in this series. New Bedford has been in the habit of sending two vessels every season to Hudson Bay, but owing to the small profits, as well as the dangerous method of prosecuting this fishery, it is more than probable that this ground will be abandoned. The vessels, usually schooners or brigs, leave their port in late spring, and after killing as many whales during the season of fishing as it is possible to do, go into winter quarters at Marble Island, where they are frozen in, and when the ice goes out make their home passages, arriving at New Bedford about September or October. When the whalemen go into winter quarters the coastal tribes build their igloos upon the ice and shores about the vessels. During the winter the Eskimo are anxious to trade, and many interesting articles of ethnological value, as well as objects of natural history, might be obtained in this manner. The whalemen— that is, the crew—trade merely for such curiosities as have an interest for them, while the vessel obtains furs and skins of land mammals.

In the season of 1881–'82, two vessels, the brig "George and Mary" and the schooner "Helen Rodman," were dispatched to Hudson Bay. The latter was wrecked shortly after her arrival. Her crew returned on the "George and Mary," which arrived October 3, 1882, and from this vessel I obtained quite a number of articles, consisting of bows and arrows, domestic utensils, and several suits of fur clothing, besides boots, shoes, and stockings, some of which are included in the series sent to London.

The returning vessel brought in part of a cargo of oil and whale-bone, and skins of the polar bear, musk-ox, and foxes.

DISPOSITION AND ARRANGEMENT OF OBJECTS.

Owing to the weight and size of some very essential objects employed in this fishery, it was decided not to send them to London, as considerable risk, delay, and inconvenience might be experienced, both in packing and in transportation; but they will be permanently installed in the National Museum. Such objects as have been selected are arranged singly and in groups, as follows: (1) Models; (2) a full sized whaleboat, with apparatus of capture; (3) upright screens, 92 by 95 inches, containing harpoons, guns, and lances; (4) a frame-work of wood, containing implements used in manipulating dead whales, blubber, and oil; (5) glass cases, containing articles of decorative art, and "scrimshaw" work peculiar to whalemen; curiosities; a series of blubber-knives; papers carried by outward-bound vessels; whalemen's journals of voyages; samples of lines and ropes used in this fishery, and accessories; and (6) a series of photographs.

1.—MODELS.

In this group are represented the whaleship, the "camels," the try-works common to all whaling vessels, and the present American whale-boat.

SHIP.—When coast-whaling was first essayed by Americans, the smaller class of vessels, such as sloops and schooners, were employed, but very short voyages were made. When, however, it was found necessary, as well as profitable, to "whale out in the deep," the smaller class of vessels gave way to barks and ships, principally the latter. These were invariably sailing vessels, until, in 1880, a bark with auxilliary steam-power, the "Mary and Helen," afterwards the "Rodgers," was successfully introduced in the North Pacific, and subsequently similar vessels owned in New Bedford and San Francisco have been sent to the same grounds. The largest fleet employed in this industry, consisting of schooners, barks, brigs, and several ships, varying from 66 to 440 tons, is owned by New Bedford. The majority of these are barks, which, as is well known, are as large as ships, the only difference being the "rig." The vessels hailing from San Francisco are principally barks, varying from 175 to 533 tons, the latter being the tonnage of the recently constructed steamer "Bowhead." The vessels owned at Provincetown, with the exception of one brig, the "D. A. Small," 119 tons, are schooner-rigged, and vary from 69 to 117 tons. Boston owns one bark of 395 tons, and several brigs and schooners of from 92 to 123 tons. Edgartown has two barks, of 301 and 314 tons, respectively, and several schooners, varying from 89 to 100 tons. New London is engaged in sealing, as well as whaling, and sends from her wharves schooners of from 134 to 250 tons. Stonington owns two schooners of 70 tons each, and Marion one or two schooners of about 84 tons.

CAMELS.—Owing to the difficulty experienced by the heavily laden whale-ships in crossing Nantucket Bar, a kind of lighter, consisting of water-tight compartments, was constructed in 1842. Since the decline of the fishery at this port the camels have been destroyed, and about the only pieces of this peculiar craft that have been saved are to be found in the garden of Mr. F. C. Sanford, of Nantucket, having been utilized in the construction of a dike or terrace.

TRY-WORKS.—The try-works peculiar to whale-ships are built of brick and mortar, framed with wood, the base resting upon the wooden sheathing of the deck. It was formerly the custom to use three try-pots, but at the present time none of the vessels have more than two. The early form of try-pot employed by Americans was manufactured in Scotland, some of which are still to be found sunning themselves about the docks at New Bedford and elsewhere, being known as the "English pot," but they are not used at present. The majority of American vessels are now fitted with try-pots manufactured at the New Bedford foundry. The largest of these weighs 1,200 pounds, with a capacity of 200 gallons; but smaller sizes of about 180 gallons are more generally used.

There is also included in this series the "head," full size, of the whaleboat, with a lay figure of the boatsteerer in the act of darting a harpoon.

2.—THE WHALEBOAT.

American whaleboats have smooth bottoms, battened seams, loggerhead aft, five thwarts, and invariably mast, mainsail, and jib. The lengths vary from twenty-eight to twenty-nine or thirty-feet. The term "craft" includes the harpoons, lances, boat-spade and boat-hook, but is oftentimes more specifically applied to the implements used to strike and kill the whale. "Boat-gear" comprehensively includes the entire outfit of the boat, but more particularly refers to the implements other than craft, such as the boat-bucket, piggin, water-bucket, line-tubs, lantern-keg, oars, paddles, and the like. It also includes the warps, but in this classification I shall mention them separately, as the main-warp or whale-line, lance-warps, short-warp and the boat-warp.

A boat's crew consists of six men; the officer of the boat, who is one of the mates, with the title of "boat-header"; the harpooner, a petty officer whose rank is next to that of a mate, known as "boat-steerer;" and five oarsmen. The boat-steerer strikes the whale, and the officer usually kills it. The oarsmen have their appointed places in the boat, and their respective duties to perform as whalemen.

3.—HARPOONS, GUNS, AND LANCES.

The implements used in the capture, pre-eminently the most important, are arranged upon the faces of four screens with maroon backgrounds, and, as far as possible, the serial and chronological order has been preserved. The first screen contains forty-seven hand-harpoons, among which may be found the forms used by the Basque, Dutch, English, French, and American fishermen, as well as a full series of the various types introduced from time to time by Americans. The second screen contains the primitive and modern types of the whaling guns, the English swivel gun, and the rocket-gun—seventeen objects in all. Upon the third screen the numerous patterns of the gun-harpoons are arranged, comprising thirty-three objects. The fourth screen is devoted to the explosive and non-explosive lances, the explosive harpoons, the rocket-bomb, seal, sea-elephant, and walrus harpoons, comprising thirty-eight implements.

These four screens may be compared to four volumes—each implement constituting a chapter—containing an exhaustive treatise on the past and present methods of the capture of the whale adopted by all nations that have participated in this fishery. The chapters, though complete in themselves, are subordinate, the subjects of the one being merely an introduction to the other, and may be used as stepping-stones as we proceed from the beginning of the seventeenth century to the present time.

HAND-HARPOONS.

The harpoons thrust by hand for striking whales may be divided into four classes: (1) the typical harpoon; (2) the common toggle-iron, and the darting-gun harpoon; (3) the hump-back iron; and (4) the prussic-acid iron.

(1) THE PRIMITIVE HARPOON.—Of this class there are properly two types: the typical harpoon with a fixed head and two barbs, and the harpoon with a fixed head and one barb. These are familiarly known as the "two-flued" and "one-flued" irons. Innovations have been made by hinging or pivoting one or two additional barbs or "flukes" in the rear of the heads of both types. None of this class are used at present by American whalemen, except possibly at times the former on the California coast, for raising "sunk" whales.

(2) THE TOGGLE-IRON.—The improved harpoon has a movable barb, known as the "toggle," pivoted at its center to the anterior end of the shank. When the instrument is to be used, the toggle is adjusted in a position parallel to the shank, and held, with the cutting point forward, by a small wooden peg. When darted into the whale the peg is broken by the resistance upon the whale line, the toggle is thrown at right angles to the shank, somewhat in the form of the letter T, and becomes transfixed in the ligamentous flesh.

The heads, toggles, or flukes, as they are also termed, may be slotted, or recessed, for the reception of the shanks; or the ends of the shanks may be slotted and the barbs pivoted between the cheeks. The latter is known as the "Temple toggle," or "Temple gig," having derived its name from the inventor, a colored man, Lewis Temple, of New Bedford, Massachusetts, who first made this kind of harpoon in about 1847 or 1848. Another mode adopted by the early manufacturers for holding the toggle in position when darted, was by means of rope, iron, or leathern grommets, which gave to the instruments the name of "grommet-irons," or "grummet-irons," as they were more frequently called. The instruments, with heads mortised for the ends of the shanks and held in position with wooden pegs, are exclusively employed by all American whalemen of the present day for fastening the whale to the boat.

To this class also belongs the present walrus-iron, which is in every particular, with the exception of size, a counterpart of the improved harpoon, and is used by whalemen in the Arctic Seas for the capture of walrus. This kind of harpoon was formerly made with a double-barbed fixed head.

Friderich Martens, in an account of a whaling voyage to the Greenland fishery during the year 1671, says: "The harpoon for a sea-horse (*Trichecus Rosmarus*, walrus or morse),* and the launce also, are short, of the length of one span, or one and a half, and an inch thick, and the wooden staff thereof is about six foot long; the harpoon for a whale is much too weak to pierce his thick skin withal, yet both of them are very well tempered and of good tough iron, and not much hardened."†

---

* Rosmarus obesus (Illig.) Gill.   † Hakluyt Society, vol. 18, p. 90.

The same author also says, in continuation of his account of the capture of the walrus, that "when they dart the harpoon at them, they always take the opportunity to do it when he is precipitating himself from the ice, or when he diveth with his head under water, for then his skin is smooth and extended, and therefore the harpoon striketh through the skin on his back the better; but when he lyeth and sleepeth, his skin is loose and wrinkled, so that the harpoon doth not pierce the skin, but falls off."

DARTING-GUN HARPOON.—The main difference between the darting-gun harpoon and the common toggle-iron is, that instead of terminating in a socket for the handle or "pole," the former has a tapering blunt point, which is intended to be placed, or, technically, "ships" into two lugs on the barrel or lock-case of the darting-gun; and also that it has a projecting iron eye, or loop, welded to the shank near the butt, into which one end of the iron-strap should be made fast. Further reference will be made to this iron in connection with the darting-gun.

(3) THE HUMPBACK-IRON.—A toggle-iron of large proportions, intended to be used only in raising sunk whales (*Megaptera* sp.) principally. Harpoons of this character are of the average length. The shanks are made of wrought iron, about 1 inch in diameter, and the heads or toggles about 10 or 10½ inches long, or almost twice the length of the ordinary toggle, and heavier. These irons are thrown into the "neck" (properly speaking the whale has no neck) or into the region about the spiracles of the humpback whale, where the blubber is exceedingly tough, as this species almost always sinks when dead. The whale remains at the bottom for two or three days, and becoming somewhat buoyant by the gases generated by incipient decomposition, it is very materially aided in making its reappearance upon the surface by the whalemen in their boats, who haul upon the large lines which are attached to the harpoons.

(4) THE PRUSSIC-ACID HARPOONS.—These harpoons were used, to a limited extent, to kill whales with prussic acid. The two instruments of this character in this series, it is supposed, were made in France and brought to Nantucket as patterns by which others might be made and introduced into the American fleet. The use of this kind of harpoon was soon abandoned, as several of the crew of a French ship were poisoned when handling the blubber of a whale killed by the acid. Although instruments of this type were carried by several American vessels, notably the ship "Susan," of Nantucket, and others, none of them, so far as the record shows, have been used, the crews having been deterred by the disastrous results experienced by the French.

POLES, STRAPS, AND SHEATHS.—One end of a rough hickory pole, oftentimes with the bark attached, is inserted into the socket of the hand-harpoon. The shank at its junction with the socket, or the socket, is served with rope yarn, to prevent iron-rust from affecting the iron-strap.

The "iron-strap," a piece of whale-line, is fastened at one end around the shank with a round turn and a splice just above the *serving*, and has an eye-splice in the other into which the tow-line is made fast.

The iron-sheaths for the heads of the instruments are made of white pine, two pieces, gouged or *scooped* out; fastened together with wooden pins, or slugs of lead, covered with canvas, and painted; usually made at sea.

WHALING-GUNS.

The guns employed in the whale-fishery were primarily intended to impel modified harpoons known as "gun-harpoons" or "gun-irons," but have been subsequently used with better success in connection with the explosive lance.

Guns of this description may be classified as (1) small arms, (2) ordnance, and (3) rocket-guns.

This classification does not embrace the so-called "harpoon-gun," which is merely an instrument with an explosive head thrust by hand, and is properly an explosive harpoon. It should be mentioned, however, that the darting-guns are sometimes known as *harpoon-guns*.

Of the first class there are two types: the single muzzle-loaders and the single breech-loaders*; the second class embraces the swivel-guns; and the third, the Roys gun and the California whaling-rocket.

SMALL ARMS.

THE SHOULDER-GUN.—The first shoulder-guns used for the capture of the whale were "muzzle-loaders," and were made with either metal or wooden stocks, and the ordinary percussion locks. Various devices have been resorted to to perfect guns of this character, some of which have not been patented. Among the most prominent may be mentioned C. C. Brand's guns with skeleton iron stocks, embracing three numbers, ranging from 1 to 3 inclusive, No. 1 being the smallest; the "Grudchos & Eggers" whaling rifle with walnut stock; the "Brown" gun with gun-metal stock and barrel; and several other kinds whose identity cannot at present be determined, among which may be mentioned those with steel barrels and walnut stocks and those with steel barrels and brass stocks.

Muzzle-loading guns were successfully employed in connection with the bomb-lances until about 1877 or 1878, at which time the improved breech-loading guns were patented and introduced. The whalemen of Provincetown, Massachusetts, prefer to use the "Brand" guns, and the whalemen of New Bedford and elsewhere invariably use the breech-loaders, which are known respectively as the "Pierce & Eggers" and the "Cunningham & Cogan." A new shoulder-gun has recently been placed on the market by H. W. Mason, of New Bedford.

The guns are discharged from the head of the boat, and are made fast

---

* Magazine guns are not used in the whale-fishery.

to the *hoisting-strap* by means of a lanyard to prevent them from being lost overboard, as the recoil of the shoulder-guns, for example, is often so great as to prostrate the gunners.

THE DARTING-GUN.—The darting-gun is a harpoon and bomb-gun combined, the former for fastening the whale to the boat, and the latter for simultaneously killing or wounding it by discharging the explosive lance, or *darting-bomb*, as it is termed. The darting-guns of the original pattern were muzzle loading, but more recent inventions have developed the breech-loaders which are known as the "screw-gun" and the "hinge gun." The whalemen recognize the two kinds in use at present as the "Pierce" and the "Cunningham," having borrowed these names from those of the inventors and manufacturers, Captain Eben Pierce and Mr. Patrick Cunningham, of New Bedford, Massachusetts.

The darting-guns are very successfully employed in all kinds of whaling, and are chiefly relied upon in the Arctic regions, where, before they were introduced, many whales escaped by running under ice after being *fastened to;* in which case, as it became necessary to cut the line to save the crew, the whale, as well as the harpoon and line, were lost.

One end of an ordinary pole, by which the apparatus is manipulated, is inserted in the rear end or socket of the gun. A harpoon is made especially for this apparatus, with a tapering blunt point which ships into the lugs on the barrel. The gun being charged and the lance inserted it is thrust by hand; the harpoon is buried in the whale, and the gun is automatically discharged by a long wire rod, which is in fact a trigger, extending beyond the muzzle, and which by impact operates the internal mechanism and projects the lance. The apparatus having been darted the whale starts off with the harpoon and exploded lance, and the gun may be hauled into the boat by a small rope and used in discharging other lances.

THE SWIVEL-GUN.—The swivel gun is of English origin, and was invented, according to Scoresby, in the year 1731, and used, it seems, by some individuals with success. Being, however, difficult and somewhat dangerous in its application, it was laid aside for many years. In 1771 or 1772 a new one was produced for the Society of Arts, which differed so materially from the instrument before in use that it was received as an original invention. This society took a great interest in promoting its introduction, and with some difficulty and great expense effected it.[*]

This kind of gun has been used by the English and Scotch whalemen in the Greenland fishery and elsewhere. American whalemen have also used the English gun, but principally in "devil fishing" and "humpbacking," in the bays and lagoons of California, "humpbacking" on the southern coast of Africa, "bowheading" in the Ochotsk sea, and in other localities where the fishery is prosecuted on soundings. Capt. John Heppingstone, of South Yarmouth, Massachusetts, tells me that

---

[*]Arctic Regions, vol. ii.

the first guns made by Capt. Robert Brown, of New London, Connecticut, were made of iron and mounted on swivels. This is the first swivel-gun, of which I have any information, manufacturd in America, with the exception of the present Mason gun.

H. W. Mason and Patrick Cunningham, of New Bedford, Massachusetts, have recently constructed a breech-loading swivel-gun, cartridge inserted in the breech, and the harpoon bomb (56376) in the muzzle, which is to be mounted in such a manner that the effects of the recoil of the gun upon the boat will be neutralized by rubber cushion-springs, for which letters-patent were issued December 12, 1882 [No. 269080, U. S. Patent Office]. Owing to the recent date of this invention, very little can be said of it, except that one of these guns has been used very successfully in the Arctic regions, and that others are being manufactured for the same fishery.

The early Dutch whalemen also used a gun with a flint lock and bell-shaped muzzle, a kind of blunderbuss, which was mounted on a swivel, notwithstanding it was provided with a wooden stock similar to that of the shoulder-gun. The first English guns were also provided with flint locks.

THE ROCKET-GUN.—The rocket-gun is of recent invention; it is supported by an iron standard, and fired while resting on, and not against, the shoulder of the gunner. It throws a large rocket and explosive lance weighing eighteen or twenty pounds, which acts in the capacity of a harpoon and bomb, and is used mainly in coast whaling or on soundings.

The rocket-gun was patented January 22, 1861, by Thomas W. Roys, of Southampton, New York, from which the California whaling-rocket is an outgrowth. Mr. C. D. Voy, of California, tells me that it was used, as far as the apparatus was concerned, very successfully on the steamer "Daisy Whitelaw," and also on the "Rocket" off the California Heads; but owing to the scarcity of whales (finbacks) in that locality, the enterprise was a failure. Mr. —— Wilson, of Sitka, Alaska, tells me that it is also used successfully, from the deck of a small steamer, by the Northwest Whaling Company in the capture of finbacks and humpbacks on the southern coast of Alaska.

GUN-HARPOONS.—The harpoons intended to be projected from guns, technically known as "gun-irons," may be used in connection with the shoulder-guns or with the swivel-guns. The shoulder-gun irons are seldom used, as the weight of the whale-line has a tendency to deflect the instrument from a true course of flight. The swivel-gun irons are employed on soundings, the heavy charge of the gun at short range overcoming the difficulty just mentioned.

Harpoons of this class may be made with double shanks joined at either end with adjustable loops composed of several wires so deftly intertwined as to conceal the ends, or of rope into which one end of the iron strap (rope) is made fast; they may be made with single shanks and sliding iron collars with rigid eyes, into which the iron straps are made fast, which, as is the case with the loops when the irons are placed in the barrel of the guns, remain on the outside; or they may be made with fluted shanks and the iron strap folded in the grooves and placed in the barrel with the instrument, the ends of the straps, to which the whale line is made fast, hanging from the muzzle.

From the following account of this kind of instrument contained in Falconer's Marine Dictionary (1830) it appears that the English at that date used a *chain strap* instead of rope for *making fast* the whale line. "Gun harpoon (*harpoon qui se darde dans un mousqueton*, Fr.), a weapon used for the same purpose as above [the Harpoon, Harping-iron, Harpon, *à pêcher les baleines*], but is fired out of a gun instead of being thrown by hand. It is made of steel and has a chain attached to it, to which the line is fastened."

The shoulder-gun irons are lighter and usually shorter than those intended for the swivel gun, and are almost always made with a movable barb or toggle; those intended for the swivel gun, though the "toggle" is the prevailing style, are sometimes manufactured with fixed double-barbed heads.

WHALEMAN'S LANCES.

The lances used in the whale fishery may be divided into two classes: (1) the non-explosive and (2) the explosive.

Of the first class there are several types, including those which may be used as hand instruments or as projectiles from guns; and of the second class many styles have been introduced which were designed to be used exclusively with guns. For convenience' sake, and in order that a more intelligent classification may be made, and a less complicated system adopted, the whale lances will be provisionally grouped as follows: (1) The non-explosive hand-lance, (2) the explosive hand-lance, (3) the non-explosive gun-lance, and (4) the explosive gun-lance or bomb.

HAND-LANCES.

THE NON-EXPLOSIVE HAND-LANCE.—The hand-lance with non-explosive head was the primitive instrument adopted by civilized races for killing whales after they had been *fastened to* with the harpoon and line. The shanks of these instruments are manufactured from the best Swedish iron, and, including the heads, vary in length from five and a half to six feet. The heads, cast-steel, are about three inches long and two inches wide, spoon-shaped, convex on both sides, and in some instances have grooves or longitudinal furrows which were probably designed, after the manner of some of the Indian arrows, to permit the egress of blood in order that it might flow freely from the wound and weaken the victim. The heads of the hand-lances have four cutting edges, and are, of course, barbless, as it is intended that the instrument should cut its way both in and out of the flesh. This instrument, which has been superseded by the bomb-lance, was always manipulated by the officer of the boat. The bow oarsman, by means of the main warp as well as by main strength, hauled the boat alongside the running whale, and the officer thrust the lance into the region of the heart and lungs, called the "life," of the cetacean, and by up and down motions, known as "churning," inflicted the mortal wound.

Notwithstanding that the explosive lance has practically done away with the use of the hand-lance, three of these instruments are at present always included in the outfit of a whale-boat, to be used in cases of emergency.

In this class should be mentioned the "fluke-lance" (56358), an illegitimate offspring of the thick boat-spade and the hand-lance, which was devised to take the place of the former during the dangerous process of "spading flukes," for stopping a running whale, in order that the boat may be hauled alongside the animal and an opportunity afforded for killing it with the hand-lance. I have been able to obtain only one example of the fluke-spade, which owes its origin to the fancy of a whaleman, and is regarded as a monstrosity by all the fraternity.

The seal lances, which may also be employed in killing the sea-elephant and walrus, but never used in whaling, on account of the short shanks, should also be grouped under this head. Such instruments have heads of varying sizes, and the ordinary shanks which terminate at the rear in sockets for the poles. They are thrust by hand, and are employed at present.

Friderich Martens, in his account of a whaling voyage to Spitzbergen, in 1671, describes as follows the method adopted by the early Dutch whalemen for the capture of the sea-horse, or sea-morse: "When great multitudes of them lie upon a sheet of ice, and they do awake and fling themselves into the sea, you must keep off your boat at a distance from the ice until the greater part of them are got off; for else they would jump into the boat to you and overset it, whereof many instances have been; then the harpoonier runs after them on the ice, or he darts his harpoon out of the boat at the sea-horse, who runs on a little until he is tired; then the men draw on the rope or line again and fetch him to the boat, where he begins to resist to the utmost, biting and jumping out of the water, and the harpoonier runs his launce into him until he is killed."[*]

THE HAND-LANCE WITH EXPLOSIVE HEAD.—The hand-lance with a non-explosive head remained for nearly two centuries the solitary type of this kind of whaling apparatus, technically known to the whalemen as "craft." On March 26, 1878, Daniel Kelleher, of New Bedford, Massachusetts, received letters-patent for an instrument, to be used as a hand-lance, which, being operated by a mechanical device coming in contact with the blackskin of the whale, should automatically explode the mag-

---

[*] Hakluyt Society, vol. 18, p. 89.

azine and imbed the fragments in the most vulnerable parts of the internal structure of the animal.

BOMB-HARPOONS.—The bomb-harpoons, or harpoons with explosive heads, also known as "harpoon-guns," of which there are two examples in this series (42762 and 56370), have detachable lance-like heads, which are chambered to receive the charge of powder, and the ordinary harpoon shank and socket. When used they are attached to poles, and thrust by hand, serving the double purpose of "fastening on to" and killing or seriously wounding the whale. Although these instruments are undoubtedly very effective, they are not regarded with much favor by the whalemen, who aver that they "are afraid of them."

One of the Provincetown schooners, when on a whaling voyage, "doubled the cape" with a box of bomb-harpoons stowed in her run; but the box was never unpacked—as the captain was unwilling to run the risk of lowering his boats with its contents—until some time after the return of the vessel to her home port, when I found the box in an old loft, and sent one of the instruments (56,370) to the National Museum. Upon its arrival the head was "soaked" in kerosene and the powder removed.

Owing to the prejudices of whalemen, these instruments have never been fairly tested, and few of the whalemen know anything of them by practical experience.

GUN-LANCES.

The lances, which are discharged from the different kinds of guns, and used with better effect and at a safer distance from the whales than were the hand-lances, are explosive and non-explosive; the former are by far the most effective, and are universally used in preference to the latter, which, although they were the results of American genius exerted in revolutionizing the system of whaling, are seldom met with in the American fishery, though worthy of mention in this class.

THE NON-EXPLOSIVE GUN-LANCE.—Subsequent to the introduction of the whaling-gun, various efforts were made to perfect a projectile for killing whales. The result was the non-explosive lance and the bomb-lance. The former has never been successfully employed. Among the most prominent of this type is the one made by Captain Josiah Ghenn, of Provincetown, Massachusetts, which was used principally for "waifing" dead whales; the one made and patented by Robert Brown, of New London, Connecticut, and several other patents of which very little is known. This kind of instrument has been supplanted by the bomb-lance.

EXPLOSIVE GUN-LANCES.—Of the explosive gun-lances there are properly four types: (1) The primitive bomb-lance for killing whales, and its modified successors, of which latter there are many kinds and which shall be designated here as *bomb-lances* to distinguish them from the following; (2) the rocket-bomb, which was invented expressly for the rocket-gun, and pre-eminently the most deadly missile that has ever been constructed for the capture of the whale; (3) the darting-bomb; and (4) the bomb-lance harpoon.

The Bomb-Lance.—The first bomb-lance patented in the United States for killing whales was invented by Oliver Allen, of Norwich, Connecticut, and is recorded in the United States Patent Office (No. 4764), under date of September 19, 1846. This instrument, unlike those which have been subsequently devised, was constructed without guiding-wings, and with an unnecessarily long tubular shank or shaft, in which was inclosed the fuse that penetrated the magazine near the anterior end of the instrument. Mr. C. C. Brand, of Norwich, Connecticut, made improvements in the Allen lance, and was, in his day, the most successful and energetic agent in developing and introducing this new mode of capturing the whale. At the death of Mr. Brand, his son, Mr. Junius A. Brand, to whom the genius of the father was transmitted, perfected the Brand lances, which are now used by all classes of whalemen. In the mean time numerous devices were constructed and patented, many of which live only in name. The evolution of this kind of lance has resulted in the "Brand," the "Pierce," and the "Cunningham & Cogan" lances, which, standing upon their special merits, are the standard lances of the age, and are to be met with in all parts of the globe where the whale fishery is prosecuted.

Although the systems of manufacturing the present lances are for the most part based upon patents recently issued, yet the inventions date back, respectively, as follows: C. C. Brand, June 22, 1852; Eben Pierce, June 1, 1869; Cunningham & Cogan, December 28, 1875, and Junius A. Brand, November 25, 1879. The term "new model," employed in the individual references to the Brand lances, is applied to those constructed at present under the patents of Junius A. Brand to distinguish them from the "old models" formerly made under the C. C. Brand system.

The magazines, or shells, of the Brand lances are cast iron, annealed, cast with heads or points which have three cutting edges, and resemble in appearance an "engraver's scraper." This lance is exploded by a time-fuse ignited by the detonation of a primer, to which fire is communicated by a firing-pin; the latter being operated upon by the discharge of the gun. The wings are of vulcanized rubber.

The shell or chamber of the Pierce lance is composed of seamless brass-tubing; the instrument has metal wings; the internal operative mechanism for exploding the lance is placed in or near the anterior end, and the explosion is caused by the concussion of the discharge of the gun, which ignites a time-fuse by means of a percussion cap.

The Cunningham & Cogan lance is composed of iron piping, to which is affixed (screwed) a malleable cast-iron point with three cutting edges. The instrument has rubber wings, and is exploded by a time-fuse ignited by a central-fire cartridge rigidly fixed to the lance and forming a part of it.

The above patents differ in their internal construction and arrangement; and, with the exception of the Brand No. 4, which is especially designed for Greener's swivel-gun, they may be used in connection with the shoulder-guns.

The Allen lance prevented the egress of water by the issue of flame in its rear caused by the burning of the fuse; the present lances are rendered impervious to water, either by tight screw-joints or by being hermetically sealed.

Pierce's and Cunningham's lances weigh, each, one and a quarter pounds, and the Brand No. 2 (new model, for example) two pounds. These weights do not include the amount of powder required for the charges.

The Rocket-Bomb.—The bomb which was designed especially for the Roys' gun, is propelled by a rocket affixed to its rear, and is the sole representative of its kind, so far as the American fishery is concerned. Further reference will be subsequently made to this projectile.

The Darting-Bomb.—The darting-bombs are short, wingless lances, made for the darting-guns, patented and manufactured by Captain Eben Pierce, Patrick Cunningham, and Mr. Junius A. Brand, respectively, and known as the "Pierce darting-bomb," "Brand darting-bomb," and the "Cunningham darting-bomb."

Bomb-Lance Harpoon.—Projectiles discharged from guns consisting of a bomb and harpoon combined have met with little success. Such instruments are intended to fasten to a whale and at the same time kill or disable it. Owing to the weight of a combination of this nature, which is unavoidable in its peculiar construction, it cannot be used in connection with shoulder-guns, as it would be impossible for man to withstand the shock of the recoil. In addition to this, the weight or "drag" of the whale-line, which must of necessity be attached, deflects the projectile from its true course, and it consequently fails to strike the whale. A harpoon of this nature, however, has recently been introduced which bids fair to overcome the obstacles just mentioned. This instrument is intended to be fired from an improved swivel-gun, and was designed by H. W. Mason and Patrick Cunningham, of New Bedford, Massachusetts, and is mentioned in the specification forming part of letters-patent (269080, United States Patent Office) dated December 12, 1882. An example of this projectile (United States National Museum, No. 56366) is included in this exhibit, and the success of the contrivance will, undoubtedly, in a short time be made known through its introduction into the fishery of the Arctic regions.

4.—IMPLEMENTS USED EXCLUSIVELY ON THE VESSEL.

The various implements employed in *cutting-in* the whale, and in mincing and boiling the blubber, are grouped upon and about a pyramidal frame-work of wood, from the center of which the immense blubber tackle lashed to mast-head shackles is suspended. The necessary chains and toggles for *fluking* the whale and for hoisting in the blubber, head, case, etc., are placed about the front. A forward cutting stage is suspended at the right, upon which a lay figure, life size, of the second mate stands with a wide cutting-spade in its hands in the act of *scarfing* the blubber. A semicircular rack in the rear contains full-sized cutting-spades of all kinds, including the heavy head-spades and the throat-spade. The case-bucket, boarding-knives, hand and machine knives for mincing blubber are displayed at the ends. The blubber-gaffs, pikes, and forks are arranged in a small rack in front, upon the left. These, together with the bailers and scrap-dippers which are in the rear, where length of space may be obtained, constitute a fair representation of the implements employed when *boiling-out*. About the top of this immense structure of whaling apparatus, which is strongly suggestive of the odor peculiar to a whaling vessel, the boat-waifs for locating dead whales are placed in prominent positions. Slabs of whalebone cross each other near the top. The superstructure consists of a main royal pole to which lookout bows are shackled. An American ensign, saturated in oil, carried by the schooner "Abbie Bradford" twelve years in the Hudson Bay whale-fishery, floats from the pole, and at the lookout a petty officer stands with a marine glass at his eye, sweeping the horizon for whales. This display contains sixty-eight objects.

CHAINS.

The chains used when working about a dead whale are the "fluke-chain," the "fin-chain," and the "head-chain." These large heavy chains are employed in the order stated: (1) for fastening the whale to the ship; (2) for raising the first "piece" of blubber with either the larboard or starboard fin, according to the side on which the whale is lying, and (3) for hoisting in the head. These chains have large triangular loose links at one end, fitting the broad thread of the blubber-hook, to prevent the strain from bursting the links, or, as they are commonly called, the "rings," although they have the form of an isosceles triangle.

THE FLUKE-CHAIN.—A large rope, known as the fluke-rope, was formerly used for *fluking* a whale, and is used now, to a limited extent; but, on the majority of the whaling vessels the chain is preferred. The process of fluking a whale, especially in rugged weather, or at night, is often accompanied by vexatious annoyances and delays. One end of the chain, with the large link, is passed around the small of the whale by means of a large buoy and rope, or by an instrument recently introduced, known as the "fluker" (55817); the other end, with the smaller link, is rove through the large link, which is slacked to the whale; the free end is taken inboard, and when the chain has been hauled taut, it is made fast to a bitt in the deck. The chain may be veered out or hauled taut as the circumstances attendant upon the cutting may require.

THE FIN-CHAIN.—The fin-chain and fluke-chain are similar in appearance, but differ in length and weight, and in the fact that the fin-chain has a large link near the middle which is used, as it is termed, for "shortening up," in order that a "longer heave" may be obtained before "coming two blocks." The fin-chain may be made without the middle loose link or ring; but those with such a ring are to be preferred. The whale having been *fluked*, the process of cutting is initiated by passing the end of the chain with the small ring around the fin, by means of a rope which is made fast to the ring. The rope and chain are then rove through the large ring, which is slacked to the fin. The blubber-hook is attached either to the middle ring or end ring; the officer cuts through the blubber around the fin, and across the whale abaft the head. The men *heave* away on the windlass, and both blubber and fin are hoisted "two blocks."

THE HEAD-CHAIN.—The head-chain, or "head-strap," as it is more frequently called, is an endless chain, with smaller links than those of the two preceding chains. It is used in right-whaling and bowheading for hoisting in the "head" (upper portion of the head) and baleen; in sperm-whaling, for hoisting in the "head," which is, in this instance, that portion consisting of the "case" and the "junk." If the whale is small, the entire "head" (junk and case) may be hoisted in; if large, these parts are taken separately. Hence, we have the apparently conflicting terms which are indiscriminately applied to this chain, namely, the "head-strap," the "case-strap," and the "junk-strap," as well as "head chain."

WHALEMAN'S SPADES.

Instruments of this character denominated "spades" by whalemen have nothing in common with the agricultural implement of the same name. In making a comparison, they may be said to resemble more nearly the common chisels used by carpenters, both in appearance, so far as the heads or blades are concerned, and in the manner in which they are used. These implements preserve their identity with remarkable accuracy. The narrow spade for "scarfing" has the same characteristics on all whaling vessels, and the same may be said of the other spades.

Of this class, used at the present time by all whaling vessels, there are (1) the "cutting-in spades," which include the "head-spade," the "throat-spade," and properly the "deck-spade;" (2) the "blubber-room spade;" (3) the "pot-spade," and (4) the "boat-spade," which, though mentioned last, was at one time an instrument of the greatest importance in capturing a whale.

The heads of these spades are made at some of the whaling ports—

principally at New Bedford—by blacksmiths who are engaged almost wholly in the manufacture of such "craft," including harpoons, hand-lances, etc. The best cast steel is used for the heads, and wrought iron for the sockets and shanks. About thirty cutting-spades without poles are included in the outfit of a whaling vessel. The poles, which are made of spruce, from fifteen to twenty feet long, are "rigged" at sea.

CUTTING-IN SPADES.—The cutting-in spades include the narrow spade for "scarfing," which is a term for cutting the blubber into spiral strips as it is unwound, or peeled, from the body of the whale; the wide cutting-spade for "leaning," severing the small pieces of flesh and muscles which adhere to the blubber; the head-spade for cutting the bone in decapitating a whale; the "sliver-spade" for detaching the pieces of flesh and blubber which connect the head and body when cutting off the head; the "throat-spade" for making a passage through the blubber of the head for the *head-strap*, and for taking out the baleen which remains in the throats of the right-whales; and the "deck-spade" for reducing to small sections the large blanket-pieces which may possibly, during the process of boarding, have to be temporarily placed on deck, before they can be lowered down the main hatch.

The above spades are used by the officers, sometimes the captain, but usually the mate and the second mate, who stand upon stages *slung* over the side of the vessel.

THE BLUBBER-ROOM SPADE.—The blubber-room spade, with a wide blade and short handle, is used between decks for the reduction of the large blanket-pieces to smaller sections, known as "horse-pieces," which are pitched upon deck, minced, and thrown into the try-works.

THE POT-SPADE.—The pot-spade is similar to the deck-spade, with the exception of the handle, which must, of necessity, be longer, as this instrument is used about the seething cauldrons of oil, for *spading* the pots to prevent the scrap from burning on the sides and bottoms and discoloring the oil.

THE BOAT-SPADE.—The boat-spade is a small, thick-set, gigantic chisel, with chamfered edges and sides, and always included in the outfit of a whale-boat, though seldom used by modern whalemen. It was mainly relied upon by the early whale-fishermen for "stopping a running whale," a process commonly known as "spading flukes." For this purpose the boat was propelled to the junction of the caudal-fin and the body—the "small" of the whale—and the animal disabled by disconnecting the cords, or by *spading* a large vein which underlies the "small"; a feat which required considerable skill and bravery, and was the most dangerous in the fishery. The introduction of the bomb-lances, however, has done away with this performance, and the whales are "stopped" as effectually at a greater distance. This spade is always carried in the boat, and used for making holes in the lips of the whale for reeving the tow-rope.

ROPES USED BY WHALEMEN.

It is not intended that the ropes exhibited in this series should include the cordage employed in the rigging of a vessel, but simply those which are used in "working about a whale," dead or alive, such as the whale-line and lance-warps used during the capture; the fluke-rope, cutting-falls, and guys, used while stripping off and hoisting in the blubber; and bone-yarn, for tying up bundles of baleen.

Whale-lines are manufactured from the best grades of Manila, loose laid, pliable, capable of bearing immense strains, and free from tar. The raw material is sprinkled with right-whale oil, during the process of dressing, to prevent the lines from rotting when exposed to salt water.

WHALEMAN'S HOOKS.

Hooks employed in the whale-fishery may be used for handling lines, chains, and blubber.

The line-hook may be used from the vessel for catching stray lines or any object afloat, but chiefly when the boat comes alongside the vessel with a dead whale, for hauling on board one end of the tow-rope, in order that the whale may be "fluked."

The large boat-hook is used from the stage, when "cutting-in," for detaching pieces of whale-line fastened to the harpoons which have been thrown into the whale during the capture, &c.

The common boat-hook is used in the whale-boat, as is any other hook of this character.

The large ring boat-hook belongs to the "cutting-gear" of the vessel, and is used from the stage, when cutting-in, for pressing upon the *back* of the blubber-hook and directing the point into the hole made in the blubber; for adjusting the fin-chain, and for hauling large pieces of blubber about deck.

The blubber-hook proper is the large hook, weighing from seventy-five to one hundred and fifty pounds, attached to the blubber-tackle and used in hoisting in the blubber.

The fin-chain hook and the small blubber-hook, or lip-hook, will be fully discussed in the subsequent individual references.

BLUBBER-TOGGLES.

The "throat-chain toggle," formerly used for taking in the throat, is essentially a *toggle*, notwithstanding the hermaphroditic sense in which the term is used. It consists of an iron toggle about eighteen inches or two feet long, and with a diameter of about three inches, with an iron strap welded around its center, forming a rigid eye, into which the "tail" or chain is made fast, and a stiff eye at one end which is used for binding or *thrap-lashing* the apparatus when toggled in the blubber. The free end of the chain has the regulation triangular link.

The common blubber-toggle is made of hard wood, and is about two feet long and six or eight inches in diameter. The wooden toggles have been used for many years for *boarding* the blubber, and are still preferred, since such an implement cannot be broken as readily as an iron toggle, especially when affected by the action of the frost. This kind of toggle, or "blubber-fid," is used in connection with the cutting-tackle, when the lower block is strapped with rope, and is, in appearance, ungainly and insignificant, but withal an important agent in the whale-fishery. A hole having been cut or *mortised* in the blubber near the fin, the eye of the block-strap (the *purchase-strop* of the English—*Admiral Smyth*) is pushed through and toggled with the fid, and the blubber hoisted in, the toggle being alternately shifted as the sections of blubber are cut from the main-piece, and lowered down the main-hatch.

WHALEMAN'S KNIVES.

It is the intention to discuss here only the knives used in connection with the blubber, which, comprehensively, may be termed *blubber-knives*. Of these the "boarding-knife," the "leaning-knife," and the "mincing-knife" are the most prominent, and are used in the order named when manipulating the blubber. Next in importance are the sheath-knives worn by the foremast hands at all times, and by the officers when *down for whales*, and the boat-knives. The former are so well known as not to require special mention here; the latter are always carried in the boats to be used in cutting the whale-line provided it "nulls" when fast to a whale, and for other purposes.

THE BOARDING-KNIFE.—Whalemen, as well as blacksmiths ashore who manufacture whale "*craft*," pick up and preserve all kinds of knives, especially those with long blades, that may be utilized either ashore or afloat in making boarding-knives. The cavalry saber and the navy cutlass are especially well adapted for the blades of this kind of knife, and are frequently used for the purpose. The whalemen visiting foreign ports also obtain by "trade," or otherwise, various kinds of knives, some of which are comparatively little known in this country. Some of them are brought home as curiosities, but others as material for the blades of boarding-knives, or for the construction of other instruments. They are, however, rarely seen in the interior, as they may be "shipped" on another voyage either in the fore-hold or in the run of the vessel, or as blades of boarding-knives; they may be consigned to the lofts where all kinds of cutting-gear are stowed, and remain for ages, or they may be lost in the mighty current which sweeps through the junk-shops, carrying with it thousands of tons of worthless material, as well as some valuable and interesting specimens which should be preserved. The "macheta" knife, so well known in tropical South America, which the natives use with such remarkable dexterity both in felling trees and in *carving* one another, frequently finds its way to the whaling-ports of this country. This kind of knife, an example of which is included in this series, is known to the Provincetown whalemen as the "cane-knife," and is used, I am told, by natives of the West Indies for cutting sugar-cane; but it is not so well adapted for the manufacture of boarding-knives as are the saber and cutlass, and is simply mentioned as a specimen of the knives sometimes found on whaling-vessels.

The boarding-knife is used by one of the officers of the vessel, usually the third mate, during the process of "boarding" the blubber, for cutting the holes, by making longitudinal thrusts through the immense blanket-piece, into which the blubber-tackle is made fast. This having been accomplished, the blanket-piece is unwound from the whale until its upper end or "head" is hoisted to the main-top, or "two blocks." The officer with his formidable-looking boarding-knife cuts off, near the planksheer, a section of the blubber, about 14 feet long, 6 feet wide, and as thick as nature has made it. This section is lowered into the blubber-room, where it is stowed away, and subsequently "leaned."

LEANING-KNIFE.—The leaning-knife resembles the ordinary butchers' knife of medium size, or the common kitchen knife, and is used in the blubber-room for "leaning blubber," that is, detaching small pieces of flesh or muscles which cling to the fat when cut from the whale, and which otherwise would *blacken* the oil when *boiled-out*.

MINCING-KNIVES.—The mincing-knives may be used, as it is termed, "by hand," or in connection with a machine designed expressly for mincing or slicing the blubber. Although these knives are used for the same purpose, yet they differ in form, and will be treated separately.

Hand Mincing-Knife.—Mincing by hand was the first method adopted and is largely in use at the present time, notwithstanding labor-saving machines have been constructed for the purpose. Hand mincing is extremely laborious, but some of the whalemen prefer this way of preparing the blubber for the try-pots, claiming that the horse-pieces are minced more uniformly, and that the oil, in consequence, is more freely boiled out. The horse-pieces are laid upon a rudely constructed bench called the "mincing-horse," and cut into slices varying from one-fourth to three-fourths of an inch thick. These slices are called "bibles" or "books"; they are not detached at the base of the piece, but are held together as are the leaves of a book, and resemble an enormous piece of fat pork. In this condition they are pitched into the try-pot.

The Mincing-Machine Knives.—This sort of knife, without handles, is rigidly fastened to a frame on the machine, and is automatically manipulated by the revolutions of a crank. The shapes of such knives vary, depending upon the kind of machine for which they are especially designed. The work of mincing is more rapidly accomplished with the machine than with the hand-knife. It is not always practicable to use the machine, owing to the yielding nature of the blubber of some species of whales, and therefore the hand-mincing knives are always carried, though the machine is included in the outfit.

5.—GLASS CASES CONTAINING CURIOSITIES AND SCRIMSHAW WORK, PAPERS AND LOGS, WHALE-LINE, AND ACCESSORIES.

These cases contain the "pans" (posterior portions of the jaw-bone of the sperm-whale, *Physeter macrocephalus*), the teeth of the same species, and tusks of walrus, engraved and carved in an artistic manner by the whalemen, as well as sundry articles manufactured from ivory and bone. Other cases are devoted to a class of objects known as "curios," brought home by whalemen from different parts of the world, including implements made and used by Eskimos of Hudson Bay, from the islands of the South Pacific, and elsewhere. This series also includes lines and ropes manufactured by the New Bedford Cordage Company, journals of voyages, copies of papers carried by the bark "Gosnold," of New Bedford, and other objects of minor importance.

6.—PHOTOGRAPHS.

A series of photographs has been made at New Bedford of whaling-vessels, docks, buildings, and whalemen. The American whaleman is represented by a group composed of both active and retired whaling-masters.* Other groups illustrative of the foreign element employed in this fishery, consist of Kanakas, Portuguese of Cape Verde, negroes from an island on the coast of Africa, and from Virginia (the latter an immense man over six feet three inches tall), Chilmark Indians from Gay Head, Massachusetts, West Indiamen, and a group of Hudson Bay whalemen attired in their fur clothing. There are also photographs of the residences of retired whaling captains, including the houses of Captain H. W. Seabury, of New Bedford, and Captain James V. Cox, of Fairhaven, and photographs of the Mariners' Home, a charitable institution where unfortunate whalemen are entertained temporarily, and the Seamen's Bethel, a place of worship erected especially for whalemen. These photographs have been enlarged by electric light, and mounted on frames thirty by forty inches.

The Species of Whales from a Commercial Standpoint.

In the ninth century, when Ohthere made his famous voyage in the Northern seas—the first record we have of killing the whale—it is believed that his captures consisted of the smaller species of cetaceans, probably of the family *Delphinidæ*, though we have no positive knowledge of the fact.

Markham† states that the Basque fishermen captured a baleen whale

---

*Group of whaling-masters of New Bedford, photographed September 14, 1882. Isaiah West, ship "Florida;" H. W. Seabury, ship "Coral;" L. Braley, schooner "William Wilson;" M. W. Taber, ship "Trident;" J. H. Cornell, ship "Janus;" Amos C. Baker, bark "A. R. Tucker;" James V. Cox, bark "Draco;" and James A. Crowell, bark "Arnolda."

† Clements R. Markham, C. B., F. R. S., paper read at the Zoological Society December 13. Published in Nature. Littell's Living Age, No. 1972, April 18, 1882.

(*Balæna biscayensis*) which had frequented their coast from time immemorial; but that this species had become nearly extinct in the seventeenth century, and that the last capture made by the Basque fishermen was on February 11, 1878, when a whale appeared off the coast between Guetaria and Zarauz. In the early part of the seventeenth century the English, Dutch, and several other contemporary European nations devoted their attention to the "whale," or Greenland whale, known to the scientific men of that age as *Balæna mysticetus*, a species of great commercial value on account of its oil and baleen. These early whalemen also made occasional captures of the sea-horse, or morse (the common walrus, *Rosmarus obesus* (Illig.), Gill, and rarely of the *Beluga*. Nantucket, at one time the leading whaling port of the world, paid exclusive attention to the capture of the sperm whale (*Physeter macrocephalus*), whose habitat is in the warmer seas; and shortly afterwards England sent vessels to engage in this fishery. "The sperm whale or nothing," seems to have been the motto of Nantucket, as none of her vessels would lower their boats for the right whale until it was too late to rectify her error. New Bedford also inaugurated her fishery on the same plan of operation, but since the decrease in value of sperm oil her vessels have willingly captured the two species of the right whale (*Eubalæna cullamach* (Chamisso) Cope, of the Pacific, and *Eubalæna cisarctica*, Cope, of the Atlantic), and the bowhead whale (*Balæna mysticetus*, Linn.), as well as humpbacks and gray whales, of which further mention will be made. The sulphur-bottom whales (*Sibbaldius sulfureus* and *S. borealis*) are seldom captured, owing to their remarkable shyness and swiftness. The California gray whale (*Rhachianectes glaucus*), ranging from the Arctic seas to Lower California, is captured by vessels at sea, by whalemen who establish stations on the California coast, as well as by the the Indians of Cape Flattery. The humpback whales (*Megaptera versabilis*, Cope, and *M. Osphyia*, Cope), frequent all oceans and are also captured. One species of this family (*Osphyia*), occasionally appears on the Cape Cod coast, following the herring inshore, and other small fish upon which it largely feeds. The finback whale of the Pacific (*Balænoptera velifera*, Cope), like the sulphur-bottom, is remarkable for its swiftness, and is therefore difficult of capture. The two Atlantic finbacks (*Sibbaldius tectirostris* Cope, and *S. tuberosis*, Cope), frequent the Cape Cod coast at certain seasons, and are captured by shore whalemen.

As is well known, *Physeter macrocephalus*, aside from the oil found in its blubber, furnishes the spermaceti, which at one time yielded handsome financial profits. But at present the demand for this product is limited, spermaceti having been supplanted by cheaper and better substitutes. This species also affords ivory and the valuable ambergris.

The right whales, so called, are now the principal objects of pursuit. Besides their oil they yield the whalebone of commerce, which, notwithstanding the numerous substitutes that have been introduced into the market, always meets with a ready sale at remunerative prices. Of the bone-bearing whales the most profitable are the *B. mysticetus*, *E. cullamach*, and *E. cisarctica*. The former also yields a fine grade of oil, known commercially as "bowhead oil" or "arctic oil." The other species, consisting of the humpback and California grays, and finbacks, yield "short bone," which is not of so much commercial importance.

The principal grades of bone are known in market as "arctic," "northwest," "South Sea," "humpback," and "calf." The smaller pieces, which are bundled separately, are known as "cullins." According to the New Bedford Shipping List, February 6, 1883, the importation of bone from January 1 of the present year to February 5, inclusive, amounted to 138,200 pounds, against 18,700 pounds during the same length of time in 1882.

Blackfish (*Globiocephalus* sp.) are also captured for their oil, and rarely the sperm-whale porpoise (*Hyperoödon bidens*) or the "square-headed grampus" of the whalemen. The latter yields a fine grade of oil, but, owing to the difficulties attendant upon its capture, whalemen seldom encounter it. The former are taken at sea, and at times on the coast of Cape Cod. The white whale (*Delphinapterus catodon* (Linn.) Gill) is occasionally captured in the rivers flowing into Cumberland Inlet, by the New London and New Bedford whalemen.

As to the present condition and statistics of the whale fishery, I submit herewith the following paper, prepared by Mr. A. Howard Clark, Assistant, United States National Museum.

STATISTICS OF THE WHALE FISHERY.

By A. Howard Clark.

The American whale fishery is now of small importance when compared with its greatly prosperous condition of thirty or forty years ago. There is still, however, a considerable number of vessels scattered over the whaling grounds in different parts of the world, and enough energy manifested in the pursuit of whales to make the business profitable in spite of the drawbacks with which it has to contend.

Three-fourths of the fleet is owned at New Bedford, which is the headquarters of the fishery. Other places, as Provincetown, Boston, and New London, in New England, and San Francisco on the Pacific coast, have an interest in the business and meet with fair success.

The entire fleet in 1880 numbered 171 vessels, aggregating 38,637.88 tons, valued, with outfits, at $2,857,650. In this fleet there were 119 bark-rigged vessels, 7 ships, 9 brigs, and 46 schooners. Two of the barks were fitted with propellers. The largest vessel of the fleet was the steam-bark Belvidere, measuring 440.12 tons, and the smallest vessel employed in ocean whaling was the schooner Union, 66.22 tons. Most of the schooners and the smaller vessels of the other classes were employed in the Atlantic Ocean whaling, while the largest and best equipped vessels were in the Pacific and Arctic fleets. The men required for these vessels numbered 4,198, and were of many nationalities, from the native American to the natives of the Sandwich or South Pacific Islands. A large proportion of the whalemen were Azorean and Cape de Verde Portuguese. The distribution of the fleet in 1880 was as follows: Hudson Bay, 5 vessels; North and South Atlantic grounds, 111 vessels; Bering Strait, 25 vessels; Pacific Ocean, 22 vessels; in port throughout the year, 8 vessels. The ownership of the vessels was divided between the different ports as follows: Ports in Massachusetts: Boston, 6 vessels; Provincetown, 20; Marion, 2; New Bedford, 123; Dartmouth, 1; Westport 2, and Edgartown, 7. In Connecticut there were 5 vessels, hailing from New London, and in San Francisco, California, 5 vessels. The interest of San Francisco in this fishery cannot, however, be measured by the number of vessels owned there, for almost the entire Arctic fleet and vessels cruising in the North Pacific are accustomed to make San Francisco a fitting-port and the headquarters for the reshipment of oil and bone to the Atlantic coast.

The value of the products of the whaling industry in 1880 was $2,636,322; the yield included 37,614 barrels of sperm oil and 34,626 barrels of whale oil, valued at $1,723,808; 458,400 pounds of whalebone, worth $907,049; and $5,465 worth of ambergris and walrus ivory. The

Pacific Arctic Ocean grounds were the most productive, yielding oil and bone worth $1,249,990. From the Atlantic Ocean grounds oil and bone were taken worth $908,771.

Most of the vessels owned at Provincetown were of the smaller class and employed exclusively in cruising in the Atlantic Ocean. The Hudson Bay and Davis Strait grounds have always been favorite resorts for the New London fleet. New Bedford vessels are found in almost all seas, with the exception of the Indian Ocean, which has been abandoned by American whalers.

Besides the vessel fishery there is a boat or shore-whaling industry, which at times is quite profitable. The only points on the Atlantic coast where boat-whaling is carried on are at Provincetown, on Cape Cod, and, one or two points in North Carolina. On the coast of California there are several stations, manned mostly by Portuguese, and on the coasts of Washington Territory and Alaska whales are taken by the Indians and Eskimo. The principal species on the Atlantic coast is the finback whale, and on the Pacific coast the California gray whale. Neither of these whales yields bone of much value, and both furnish but a limited quantity of oil. Humpback, sulphur bottom, and right whales are occasionally taken along the California and Alaskan coasts, but seldom on the Atlantic.

The whale fishery of this country was in its zenith of prosperity about the middle of the present century, when the fleet numbered 736 vessels, aggregating 231,406 tons. From 1854 to the present time there has been an almost constant decrease in the size of the fleet. The chief cause of this decline has been the introduction of mineral and cotton-seed oils, at very low prices, which made a great reduction in the value of whale oils, and has rendered the cost of production equal to if not greater than the market value of those articles. The products of the whale fishery in 1854 were of greater value than for any year before or since, being $10,766,521, against $2,056,069 in 1879, which was the lowest value since 1828, when the fishery yielded $1,995,181. The largest quantity of sperm oil was in the year 1837, when the fleet landed 5,329,138 gallons, worth $6,650,000. The largest quantity of whale oil was in 1851, when there were landed 10,347,214 gallons, worth $4,656,000. In 1853 the amount of whalebone taken was 5,652,300 pounds, worth $1,917,000; the largest amount in any year of the history of the business. The value of bone has, however, greatly increased since that period, and is now more than of $2 per pound.

The relative importance of the various whaling grounds during the past years, from 1870 to 1880, is shown by the following facts. Of the sperm-oil landed during that period, 55 per cent. was taken in the North and South Atlantic Oceans, 33 per cent. in the Pacific, and 12 per cent. in the Indian Ocean. Of whale-oil, 58 per cent. came from the North Pacific and the Pacific fleets, 24 per cent. from the North and South Atlantic fleets, 10 per cent. from the South Pacific, 5 per cent. from the

Indian Ocean, and 3 per cent. from Hudson Bay, Cumberland Inlet, and Davis Strait. Of the whalebone secured in the same decade, 88 per cent. was from north of the fiftieth parallel in the North Pacific and Arctic Oceans, and 5 per cent. from Hudson Bay and Cumberland Inlet, and the balance from the Atlantic, Indian, and other oceans.

The number of voyages undertaken by the fleet for 1870 to 1880 was 810, which includes Arctic whalers annually refitting at San Francisco and other ports. Of these voyages 382 were to the North and South Atlantic, 254 to the Pacific, Arctic, and adjacent grounds, 98 to the South Pacific, 45 to the Indian Ocean, and 36 to the Atlantic Arctic grounds, Hudson Bay, Davis Strait, and Cumberland Inlet.

Year.	Ships and barks.	Brigs.	Schooners.	Total vessels.	Total tonnage.
1870	217	22	77	316	72,173
1871	214	18	48	280	67,900
1872	170	12	28	210	51,252
1873	151	12	28	191	46,230
1874	129	7	25	161	39,788
1875	118	8	26	152	36,230
1876	121	7	30	158	37,182
1877	120	8	34	162	36,476
1878	128	11	39	178	39,976
1879	123	12	42	177	39,391
1880	126	9	46	181	38,637

Table showing the value of sperm-oil, whale-oil, and whalebone landed by the American fleet, the value of the consumption in the United States, and the value of the exportation annually from 1870 to 1880.

Year.	Landed by the fleet.*	Consumption in the United States.	Exportation.
1870	$4,529,126	$2,896,883	$1,476,864
1871	3,691,469	2,798,408	1,479,153
1872	2,954,783	2,081,468	1,374,098
1873	2,962,106	1,947,037	929,247
1874	2,713,034	2,154,638	1,179,286
1875	3,314,800	1,700,823	1,494,727
1876	2,639,463	1,346,828	1,487,533
1877	2,300,569	1,113,681	924,175
1878	2,232,029	849,043	1,357,162
1879	2,056,069	1,345,582	582,994
1880	2,659,725	1,165,944	795,657

* From half a million to a million dollars' worth of products are carried over from year to year.

FISHERIES OF THE UNITED STATES.

Table showing the number of barrels of sperm and whale oil and pounds of whalebone landed by the American fleet, the quantities consumed in the United States, and the quantities exported annually from 1870 to 1880.

SPERM-OIL.

Year.	Amount landed.	Consumption.	Exportation.
	Barrels.	*Barrels.*	*Barrels.*
1870	55,183	28,812	22,733
1871	41,534	33,528	22,156
1872	45,201	24,052	24,344
1873	42,053	24,190	16,238
1874	32,203	21,708	18,075
1875	42,017	18,453	22,802
1876	39,811	14,473	23,600
1877	41,119	31,737	18,047
1878	43,508	11,124	32,769
1879	41,308	23,315	11,843
1880	37,614	13,750	12,283

WHALE-OIL.

1870	72,691	68,452	9,672
1871	75,152	63,011	18,141
1872	31,075	42,852	1,528
1873	40,014	33,881	2,153
1874	37,782	44,357	3,300
1875	34,594	31,800	5,424
1876	33,010	22,620	10,300
1877	27,191	20,501	6,390
1878	33,778	12,557	14,371
1879	23,334	24,685	7,374
1880	34,776	23,856	4,395

WHALEBONE.

	Pounds.	*Pounds.*	*Pounds.*
1870	708,365	255,347	347,918
1871	600,655	319,856	387,199
1872	193,793	74,141	177,932
1873	206,396	155,351	120,545
1874	345,500	208,807	165,553
1875	372,303	143,007	205,436
1876	150,628	150,628	133,400
1877	160,220	67,820	70,800
1878	207,259	96,859	113,400
1879	286,280	183,565	75,715
1880	464,028	176,770	171,258

DESCRIPTIVE LABELS ACCOMPANYING THE OBJECTS.

MODELS.

WHALE-SHIP, "CAMELS," WHALE-BOAT, AND TRY-WORKS.

MODEL OF WHALE-SHIP.
> Sails clewed up and down for cutting. Sperm-whale alongside, decapitated; forward and after stages rigged outboard. Try-works between foremast and mainmast. Four boats on the cranes; two spare boats on the skids. Officers engaged in cutting and boarding; crew at windlass. American ensign at mizzen peak. Length over all, 4 feet 4 inches; beam, 11½ inches. Edgartown, Massachusetts, 1876. 25726. C. H. Shute & Son.

CAMELS.
> Two water-tight compartments; each provided with a propeller, a smokestack, and a series of windlasses. Scale, 1 inch to 5 feet 5 inches. Length, 2 feet 1¼ inches. 25027. William H. Chase. A kind of lighter made in two sections, divided lengthwise, for floating loaded vessels over Nantucket Bar. The model with the hull of a vessel shows the manner in which the "camels" were operated.

WHALE-BOAT.
> One-sixth the length of a thirty-foot boat, from which it was drafted, illustrative of all the parts of a boat used in Arctic whaling, with mast, oars, and rowlocks. New Bedford, Massachusetts, 1883. 57199. Made by James Beetle. U. S. Fish Commission.

TRY-WORKS.
> Model of try-works common to all whaling vessels. Two pots "set"; copper cooler, wooden scrap-hopper, cast-iron deck-pot, accompanied by miniature models of the bailer, dipper, oil-scoop, and pot-spade. 17¼ by 12 by 8¾ inches. New Bedford, Massachusetts. 25013. Captain L. W. Howland.

APPARATUS OF CAPTURE.

AMERICAN WHALE-BOAT FULLY EQUIPPED FOR THE CAPTURE.

WHALE-BOAT.

A full-sized whale-boat with apparatus of capture. Length, 28 feet; beam, 5 feet 10 inches. New Bedford, Massachusetts. 72795. Gift of I. H. Bartlett & Sons.

OARS.

White ash. One steering-oar, 22 feet in length; and 5 pulling-oars, used by the boat-steerer, bow-oarsman, midship-oarsman, tub-oarsman, and stroke-oarsman. The oars for the oarsman vary in length as follows: Two about 16 feet long, two 17 feet, and one 18 feet. New Bedford, Massachusetts, 1882. Steering oar, 72796; harpoon-oar, 72797; bow-oar, 72798; midship-oar, 72799; tub-oar, 72800; stroke-oar, 72801. Gift of I. H. Bartlett & Sons.

PADDLES.

Made in two pieces, handle and blade. Sometimes used when approaching a whale in calm weather. Six paddles complete the outfit. New Bedford, Massachusetts, 1882. 72802; 72803; 72804; 72805; 72806; 72807. I. H. Bartlett & Sons.

ROWLOCKS.

Common spur oarlocks, iron, used by the boat-steerer, bow-oarsman, midship-oarsman, and stroke-oarsman. New Bedford, Massachusetts. 72821. Gift of I. H. Bartlett & Sons.

TUB-OARLOCK.

A double-decked oarlock, iron, with two rests for the oar, used by the tub-oarsman, who, when propelling the boat, uses the lower lock, and when fastened to the whale shifts the oar to the upper lock in order that the line may run out freely. New Bedford, Massachusetts, 1882. 72822. I. H. Bartlett & Sons.

HARPOONS.

Six toggle-irons; two on the iron crotch, and four spare irons. First iron attached to end of whale-line, second iron connected with the standing part of tow-line with the short warp. New Bedford, Massachusetts, 1882. 72824. I. H. Bartlett & Sons.

HAND-LANCES.

Three hand-lances, wrought-iron shanks, steel heads, wooden poles, strapped and rigged. Tied with nettles. New Bedford, Massachusetts. 72825. Gift of I. H. Bartlett & Sons.

BOAT-SPADE.
> Wrought-iron spade edged with steel, wooden pole, strapped and rigged. Formerly used for stopping a running whale by severing the tendons at the junction of the caudal fin with the body, and used at present for cutting holes in the lips of the whale for reeving the tow-rope. New Bedford, Massachusetts. 72827. Gift of I. H. Bartlett & Sons.

WHALING-GUN.
> Brand gun No. 2; muzzle-loading, skeleton iron stock. New Bedford, Massachusetts. 72820. Gift of I. H. Bartlett & Sons.

WHALEMAN'S LANCE-BAG.
> Canvass bag used as a receptacle for bomb-lances in the boat when down for whales. The lances having been placed in the bag, which is painted or tarred, to render it impervious to water the wooden stopper or plug is inserted at the mouth and tightly bound with the twine. New Bedford, Massachusetts, 1882. 72819. Gift of I. H. Bartlett & Sons.

BOAT GRAPNEL.
> An iron hook, with four barbless arms, used for picking up lines or other objects floating in the water when working about a dead whale prior to towing it to the vessel. Stiff ring for rope. Height 9 inches. New Bedford, Massachusetts, 1882. 72829. Presented by J. Barton.

IRON CROTCH.
> A two-pronged implement in which the handles of the harpoon are placed when the boat is approaching a whale. Wood. One piece scarfed in two places at the top, and filled in with wedge-shaped pieces of wood, the projecting ends forming "crotches," into which the iron poles are placed. Small iron spike inserted into the foot, which is protected by a brass ferule. The iron spike ships into a socket in a cleat nailed to the inner edge of the gunwale on the starboard bow. Fastened to the boat with a small laniard. Length, 19 inches. Laniard, 22 inches. New Bedford, Massachusetts, 1882. 72823. I. H. Bartlett & Sons.

SHORT WARP.
> A piece of whale-line fastened to the main warp with a bow-line and used to connect the second iron. Length, 4 fathoms. New Bedford, Massachusetts. 72828. Gift of I. H. Bartlett & Sons.

LINE-TUBS.
> Receptacles for the whale-line. Oak staves bound with iron hoops. Bottoms perforated with numerous holes intersected by cross-

LINE-TUBS—Continued.

grooves cut into the wood, forming outlets and channels, through which such water as may accidentally get in the tubs may escape. Semicircular cut in upper end of one stave, through which the line is paid out when fast to a whale. Two ropes are spliced in staves on opposite sides and used as lashings, with which the tubs are made fast to thwarts to prevent their loss overboard. Large, 72808. New Bedford, Massachusetts. 1882. Small, 72809. Gift of I. H. Bartlett & Sons. Two tubs: The large tub, circular, contains 225 fathoms of line; the small tub, elliptical, contains 75 fathoms.

WHALE-LINE.

Manila, slack laid, three strands, circumference 2 inches. Three hundred fathoms of whale-line are usually carried in a boat, seventy-five fathoms in the small tub and two hundred and twenty-five in the large tub. Laid in Flemish coils in order that the line may run out freely when fastened to a whale. The "top end" of the line in the large tub is led forward and fastened to the first iron, and the bight of the line thrown over the loggerhead. New Bedford, Massachusetts, 1881. Large line. 72810. Small line, 72811. Gift of I. H. Bartlett & Sons.

DRUG (DRAG).

A drug made in the form of a tub, with a thick and strongly made bottom to withstand the resistance of water. Oak staves bound with iron hoops. Upright piece of hard wood morticed and toggled in bottom. Rope tail for bending on to harpoon. Used to impede the progressive motions of a wounded adult whale, or fastened to a calf to attract the attention and sympathy of its mother or other females. New Bedford, Massachusetts, 1882. 72843. I. H. Bartlett & Sons.

LANTERN-KEG.

A utensil included in the outfit of every American whale-boat, sometimes made by the cooper on board the vessel, and sometimes ashore. Oak staves; three iron hoops. Headed up at both ends. New Bedford, Massachusetts. 72812. I. H. Bartlett & Sons. The lantern-keg contains the boat-lantern, matches, tinder box, candles, pipes, and tobacco, and sometimes ship-bread. Its position in the boat is invariably in the apartment aft, known as the cuddy, under which it is "slung" by rope-lashings.

BOAT-BUCKET.

A strongly made tub, heavy oak staves, with two projecting staves with holes in upper ends for a knotted rope bail or handle; iron hoops. New Bedford, Massachusetts. 72813. I. H.

BOAT-BUCKET—Continued.
> Bartlett & Sons. As soon as the boat is fast to a whale the order is given to "Wet line!" and the man whose duty it is grasps the boat-bucket, and, dipping water from overboard, pours it into the line-tub to prevent friction as the line runs rapidly round the loggerhead.

BOAT-PIGGIN.
> An ordinary piggin with a projecting stave as a handle, used for bailing the boat. New Bedford, Massachusetts, 1881. 72814. Gift of I. H. Bartlett & Sons.

WATER-KEG.
> Oak staves, headed at both ends, bound with iron hoops, with an outlet for water on the upper end. Used as a receptacle for fresh water for the men when down for whales. New Bedford, Massachusetts, 1881. 72815. Gift of I. H. Bartlett & Sons.

BOAT-LANTERN.
> A small oblong lantern with glass sides, and a tin socket for the reception of candles. Used as an ordinary lantern in the boat when down for whales if the capture is prolonged until night, and as a signal for the ship. New Bedford, Massachusetts, 1881. 72816. Gift of I. H. Bartlett & Sons.

BLACKFISH POKE.
> The stomach of the Blackfish (*Globiocephalus*) deprived of its inner membrane, inflated, and dried, painted white, wooden plug inserted and seized in neck. Provincetown, Massachusetts. 72844. Gift of Stephen Cook. At times, when a whale is fast to a boat, it may run so rapidly, or sound to such a depth, as to take out all the line. Under such conditions the poke is bent on to the end of the line before it leaves the boat, and when the whale ceases its progressive motions the poke or buoy appears on the surface and the line is regained. It is also used in waifing dead whales or blackfish.

BOAT-HORN.
> Used in a whale-boat as a fog-horn. Tin, japanned; mouth-piece, tin. Length, 13¼ inches. Diameter of mouth, 2¼ inches. New Bedford, Massachusetts, 1882. 72818. U. S. Fish Commission.

TINDER-BOX.
> A water-proof box carried in the lantern-keg. Tin, small ring handle, painted. Contents: Flint, steel, and cloth, for lighting pipes when down for whales, or, perhaps, making fires on shore, if the boat should be engaged in whaling on soundings, and the crew feel disposed to warm themselves or to have a hot meal. Height, 2 inches. Diameter, 4½ inches. New Bedford, Massachusetts, 1882. 72817. Presented by John McCullough.

CRAFT.

WHALEMAN'S HARPOONS.

NON-EXPLOSIVE—THRUST BY HAND.

TWO BARBS AND FIXED HEAD.

HARPOON.

Harpoon with fixed head, double barbed. Slender neck. Length, 37 inches. 25010. U. S. Fish Commission.

HARPOON.

A very old harpoon, with a double-barbed fixed head, worn out in service by frequent applications to the grindstone. Wrought iron. Eye-splice for iron strap grafted at socket. Cut from a dead whale. Length, $30\frac{1}{4}$ inches. 25902. U. S. Fish Commission.

THE CARSLEY HARPOON.

Patented by William Carsley, of New Bedford, Massachusetts, July 29, 1841. A two-flued harpoon with fixed head, so constructed that the barbs are made to stand obliquely to the axis of the shank. On entering the whale the instrument cuts "its way in an oblique or spiral direction, making the incision such that when a strain is brought to bear upon the line attached to the harpoon, either by the resistance of the animal, its efforts to escape, or otherwise, the flukes or barbs will be brought into a position more or less nearly at right angles with the lips of the incision, making it vastly more difficult than is the case with the common harpoon for it to be drawn out backward by returning in the direction of the cut or wound."—(Specification of patent.) Length, 37 inches. Fairhaven, Massachusetts, 1882. 56226. Painted vermilion. Manufactured and presented by Luther Cole.

DOUBLE-BARBED HARPOON.

Primitive style. Has evidently been used in capturing a whale. Not employed at present. Length, $31\frac{3}{4}$ inches. 56246. U. S. Fish Commission.

TWO-FLUED HARPOON.

A kind of harpoon known as the "English harpoon," formerly used in the American fleet, having been obtained from English vessels. Head fixed; two barbs with reverse barbs or "beards" on the rear ends. Chilled wrought iron. Marked "L. B. & Co." (Manufacturers). Length, $38\frac{1}{2}$ inches. New London, Connecticut. 56257. Gift of C. A. Williams & Co.

TWO-FLUED HARPOON.

A kind of harpoon with fixed head employed in the American fleet during the early days of whaling. Has been used in capturing a whale. Wrought iron. Length of harpoon, 42 inches. Length of barbs, 6¾ inches. New London, Connecticut, 1882. 56258. Gift of Lawrence & Co.

HARPOON.

Harpoon with a double-barbed fixed head; wrought iron. Marked "Macy" (Manufacturer). Length, 30 inches. New Bedford, Massachusetts, 1882. 56259. Presented by Mackey & Pindar. A style of harpoon with a large head. Formerly used.

TWO-FLUED HARPOON.

Harpoon with double barb and fixed head. Length, 29½ inches. New Bedford, Massachusetts 1882. 56263. U. S. Fish Commission.

TWO-FLUED HARPOON.

Harpoon with two fixed barbs, and one movable barb pivoted at one side at rear of fixed head. Barbs and shank wrought iron. Length, 33 inches. Provincetown, Massachusetts. 56264. U. S. Fish Commission.

TWO-FLUED HARPOON.

Harpoon with double-barbed fixed head; wrought iron. Diameter of shank reduced by tractile force. Length, 32¼ inches. New Bedford, Massachusetts, 1882. 56265. Gift of Mackey & Pindar.

TWO-FLUED HARPOON.

Common harpoon with a fixed wrought-iron head; two barbs. Shank, wrought iron, reduced in diameter by tractile force. Cut from a dead whale. Marked "B K P   T" (initials of bark). Length, 34 inches. New Bedford, Massachusetts, 1882. 56266. Gift of John A. Sawyer.

TWO-FLUED HARPOON.

Harpoon with double-barbed fixed head; wrought iron. Stamped "G. S." (Manufacturer). Length, 36 inches. Nantucket, Massachusetts, 1882. 56267. Gift of Joseph B. Macy.

WHALEMAN'S HARPOON.

The typical harpoon, with pole and iron strap, formerly used for striking whales, but superseded by the improved style technically termed the "toggle-iron." Total length, 9 feet 5 inches. New Bedford, Massachusetts. 56403. Gift of John McCullough. This harpoon is over thirty years old, and was found in the loft of a warehouse where it had remained for as many years.

FISHERIES OF THE UNITED STATES.

WHALEMAN'S HARPOON—Continued.

It belongs to that series of harpoons which have fixed heads and two barbs, and is in the condition in which it is intended to be used, with the exception that the points should be ground to a razor's edge and the shank cleaned and polished, in order that no obstruction may be offered when penetrating the blubber and flesh.

SINGLE-BARBED HARPOONS—FIXED HEADS.

ONE-FLUED HARPOON.
Harpoon with single recurved barb. Marked "S. Lydia." Length, 32¼ inches. New London, Connecticut, 1882. 56249. Gift of Lawrence & Co.

ONE-FLUED HARPOON.
Harpoon with a fixed single barb and hinged toggle. Barbs and shank, wrought iron. Length, 35 inches. Provincetown, Massachusetts, 1882. 56250. U. S. Fish Commission.

HARPOON.
Harpoon with one recurved fixed barb and one adjustable barb. The latter is rigidly fastened to the forward end of a wrought-iron rod. The rod is made fast to a cast-iron sliding collar or socket. The sliding socket has an eye into which one end of the iron strap may be bent. The combination of sliding socket, arm, and adjustable barb moves around, or parallel to, the shank. A wrought-iron eye is welded near the rear end of the shank, through which the iron strap may be rove. When the instrument is to be darted, the adjustable barb is closely fitted to the rear of the fixed barb, where it is held in position by a small wooden pin. The resistance upon the line, which is rove through the stationary eye and made fast to the rigid eye on the sliding socket, gives the under barb a twisting motion which brings it at right angles, or otherwise, to the point of incision, more firmly fastening the instrument in the flesh of the whale. Length, 35 inches. New London, Connecticut, 1882. 56251. Gift of C. A. Williams & Co. Probably a modification of a harpoon patented October 20, 1857, by James Q. Kelly, of Sag Harbor, New York.

ONE-FLUED HARPOON.
Harpoon with fixed head, single barb. Barb and shank, wrought iron. Marked "E. Park" (Manufacturer). Cut from a dead whale. Length, 36 inches. Nantucket, Massachusetts, 1882. 56252. Gift of Joseph B. Macy.

ONE-FLUED HARPOON.

Harpoon with single barb and fixed head; diameter of neck of shank reduced, in order that it may be bent in the weak place by the action of the flesh and act upon the principle of a toggle; head and shank, wrought iron. Length, 37 inches. New Bedford, Massachusetts, 1882. 56253. Manufactured by Mr. James D. Driggs and presented by Capt. James V. Cox. This style of harpoon was at one time used, and it is believed by some whalemen that it suggested the idea of the toggle-iron. None of them are used at present.

ONE-FLUED HARPOON.

Harpoon with one fixed barb and hinged toggle; barbs and shank, wrought iron. Socket, partly wrapped with marline. Marked "J. B. Morse" (Manufacturer). Length, $33\frac{1}{4}$ inches. Edgartown, Massachusetts, 1882. 56254. U. S. Fish Commission.

ONE-FLUED HARPOON.

Harpoon with fixed head; single wrought-iron recurved barb. Shank, wrought iron. Marked "Howard, 1111" (Name of vessel and number of boat). Cut from a dead whale. Length, 33 inches. New London, Connecticut, 1882. 56255. Gift of Lawrence & Co.

IMPROVED HARPOON, OR TOGGLE-IRON.

TOGGLE-IRON.

Toggle, malleable cast iron, mortised and pivoted to shank. Shank, wrought iron; slotted for toggle. Point broken. Length, $31\frac{1}{4}$ inches. New Bedford, Massachusetts, 1876. 25642. Gift of W. H. Cook & Co. Has been used in capturing a whale.

TOGGLE-IRON.

Diameter of shank reduced by tractile force. Head, malleable cast iron, mortised. Shank, wrought-iron. Cut from a dead whale. Length, 36 inches. Provincetown, Massachusetts. 29308. Presented by Capt. J. G. Fisher. Used by Captain Fisher in capturing a whale.

THE DOYLE HARPOON.

A kind of harpoon invented by George Doyle, of Provincetown, Massachusetts, November 2, 1858. The nature of this invention consists in attaching the shank to the head in such a manner that when the harpoon has been thrust into the whale it shall present the broad flat side of the head instead of the rear edge. The head may be made of cast steel or other suitable material, with a longitudinal slot extending from the center backward to receive the end of the shank. Shank, wrought iron, pivoted to the head with a steel pin. Length, $34\frac{1}{4}$ inches.

FISHERIES OF THE UNITED STATES.

THE DOYLE HARPOON—Continued.

Provincetown, Massachusetts, 1882. 56231. Gift of I. A. Small.

TOGGLE-IRON.

Shank twisted to show the tenacity and durability of the iron employed in the manufacture of harpoon-shanks. Head consists of the common malleable cast-iron toggle, with a diamond point. Shank, wrought-iron. New Bedford, Massachusetts, 1876. 56233. Gift of E. B. & F. Macy.

TOGGLE-IRON.

Shank twisted by the actions of a dying whale. Marked "D. & B." (Manufacturers); "B K D A R") (initials of bark). Length, 29½ inches. New Bedford, Massachusetts, 1882. 56237. Gift of Jonathan Bourne.

TOGGLE-IRON.

Shank elongated by tractile force. Head, malleable cast iron, mortised. Shank, wrought iron. Cut from a whale. Length, 32¼ inches. New Bedford, Massachusetts, 1882. 56242. Gift of John A. Sawyer.

TOGGLE-IRON.

A kind of harpoon, with a small toggle known as the "Sag Harbor iron." Head, malleable cast iron. Shank, wrought-iron. Length, 38½ inches. New Bedford, Massachusetts, 1882. 56243. Gift of Mackey & Pindar.

TOGGLE-IRON.

Head, wrought iron, pivoted in cheeks of forward part of shank. Toggle has two flukes, one at forward and one at after end. Shank, wrought iron; head mortised for toggle. Marked " E. S." (Manufacturer). Length, 38½ inches. New Bedford, Massachusetts, 1882. 56244. Gift of Jonathan Bourne.

NICKEL-PLATED TOGGLE-IRON.

Head, malleable cast iron, mortised; shank, wrought iron. Length, 34 inches. Fairhaven, Massachusetts, 1882. 56245. Manufactured and presented by Luther Cole. The kind of harpoon, commonly known as the "toggle-iron," used at present by all American whalemen. Full size.

TOGGLE-IRON.

Harpoon with movable head or toggle pivoted in cheeks of shank, with two barbs at point and a flaring barb or fluke at rear extremity. Shank, wrought iron, slotted for toggle. Cut from a dead whale. Length, 33 inches. New Bedford, Massachusetts, 1882. 56247. Gift of Jonathan Bourne.

TOGGLE-IRON.
> Toggle-iron with loop twisted in the shank by the actions of a wounded whale. Head, malleable cast iron; shank, wrought iron. Length, 30¾ inches. New Bedford, Massachusetts, 1882. 56248. Gift of Aiken & Swift. (Manufactured by Luther Cole, Fairhaven, Massachusetts.)

GROMMET-IRON.
> Head consists of a mortised steel toggle, with a smooth point and a barb or fluke at the rear extremity; shank, wrought iron; socket served with marline and with an eye-splice for bending on the whale-line; two rope grommets for holding the toggle in position when darted. Length, 26 inches. New Bedford, Massachusetts, 1882. 56256. Gift of Messrs. L. & W. R. Wing & Co. A kind of iron known as the "grommet-iron;" may be used in striking the walrus, but not used at present in capturing the whale.

GROMMET-IRON.
> Head, toggle with diamond point, wrought iron, held in position when darted by an iron link or "grommet;" shank, wrought iron; socket wound with marline, around which the iron-strap is fastened. Length, 26½ inches. New Bedford, Massachusetts, 1882. 56268. Gift of John McCullough. A kind of harpoon manufactured at an early date and known as the "grommet-iron" from the fact that its barb or toggle was confined to the shank with a "grommet" instead of a small wooden pin, which latter is now in general use. None of the irons of this character are used at present. They have been used in striking the whale or walrus, being better adapted for the latter on account of the short shank and small head.

TOGGLE-IRON.
> Head and portion of shank of common toggle-iron. Toggle pivoted to shank. Length, 9¾ inches. 56404. Gift of A. R. Crittenden.

TOGGLE-IRON.
> Head and portion of shank of common toggle-iron. Toggle, malleable cast iron. Length, 6¾ inches. 56405. Gift of A. R. Crittenden.

TOGGLE-IRON.
> Head and portion of shank of toggle-iron. Evidently a kind of lily-iron intended to be used for striking the sword-fish or porpoise. Toggle with double diamond point; slotted and hinged at center to end of shank. Shank, wrought iron. Length, 10½ inches. 56406. Gift of A. R. Crittenden.

TOGGLE-IRON.
Head and portion of shank of toggle-iron. Evidently intended for striking sword-fish, porpoises, and black-fish. Head, steel, mortised. Shank, wrought iron. Length, 10 inches. 56407. Gift of A. R. Crittenden.

TOGGLE-IRON.
Head and portion of shank of toggle-iron. Toggle pivoted between the cheeks of shank. Toggle has two barbs; one front and rear. Length, 9¾ inches. 56408. Gift of A. R. Crittenden.

TOGGLE-IRON.
Head and portion of shank of toggle-iron. Toggle, wrought iron, elongated point, fluked at rear end; slotted and pivoted to end of shank. Intended to be used in striking sword-fish or porpoises. Length, 10¾ inches. 56409. Gift of A. R. Crittenden.

TOGGLE-IRON.
Head and portion of shank of a harpoon known as "Doyle's iron." Patented November 2, 1858, by George Doyle, of Provincetown, Massachusetts. Head, malleable cast iron. Length, 12¾ inches. 56410. Gift of A. R. Crittenden. The peculiarity of the head of this kind of iron is that it is so arranged as to present the flat side of the blade when fastened in the whale.

TOGGLE-IRON, WITH POLE.
Toggle, malleable cast iron; shank, wrought iron. New. Length, including pole, 10 feet 4 inches. Length of harpoon, 33 inches. Length of toggle, 8 inches. Fairhaven, Massachusetts, 1883. 56416. U. S. Fish Commission. The present form of harpoon used by American whalemen.

TOGGLE-IRON, WITH POLE.
Toggle made of hoop-iron, held by a rope grommet. Roughly-made pole. Socket served and iron strap attached. Length, 9 feet. New Bedford, Massachusetts, 1882. 57698. Gift of Jonathan Bourne. A peculiar harpoon made on board ship, probably by the blacksmith, evidently intended for striking blackfish or porpoise.

HARPOONS FOR RAISING "SUNK" WHALES.

HUMPBACK-IRON.
Toggle, malleable cast iron, mortised, and pivoted to shank. Shank ⅝ inch wrought iron. Total length, 34 inches. Total length of toggle, 10½ inches. New Bedford, Massachusetts, 1882. 56269. Gift of Jonathan Bourne. A kind of harpoon, known as the "Humpback-iron," for raising sunk whales.

HUMPBACK-IRON.

Toggle, malleable cast-iron, mortised, pivoted to shank. Shank, wrought iron, with slot for toggle. Total length, 40 inches. Length of barb, 10¾ inches. New Bedford, Massachusetts, 1882. 56270. U. S. Fish Commission (manufactured by Jas. Barton). A kind of harpoon, known as the "Humpback-iron," used for raising sunk whales (*Megaptera* sp.).

HUMPBACK-IRON.

Rough specimen of a kind of harpoon, known as the *Humpback-iron*, used in raising sunk whales. Wrought iron, forged and manufactured on board ship. Length, 33 inches. New Bedford, Massachusetts, 1882. 56271. Gift of Jonathan Bourne.

DARTING-GUN HARPOONS.

THRUST BY HAND, BUT NOT FIRED FROM THE GUN (NEW).

PIERCE'S DARTING-GUN IRON.

Head, malleable cast iron; shank, wrought iron; loop welded to shank near the butt for making fast the end of the tow-line. The rear end tapers to a blunt point for insertion into the lugs of the darting-gun. Length, 29 inches. New Bedford, Massachusetts, 1876. 25680. Gift of W. H. Cook & Co.

THRUST BY HAND AND FIRED FROM THE GUN (OLD.)

PIERCE'S DARTING-GUN HARPOON.

Temple gig. Shank consists of two conjoined parts; forward part wrought iron, extreme end mortised for the toggle; rear portion, iron piping with screw at one end for tip; tip wanting; iron arm pivoted in a slot in the shank at one end; in the other end there was intended to be an eye, or loop, in which the strap for bending on the whale-line should be fastened. Eye, wanting. Length, 23¼ inches. Provincetown, Massachusetts, 1882. 56209. Gift of I. A. Small. Formerly used in connection with the first darting-gun.

PIERCE'S DARTING-GUN HARPOON.

Temple gig, shank in two sections; forward part, wrought iron, extreme end mortised for toggle; rear portion, iron piping. Wrought-iron link for making fast end of whale line. Button, felt. Length, 23 inches. Provincetown, Massachusetts, 1882. 56222. Gift of Capt. William Roberts. Made especially for the first darting gun.

PIERCE'S DARTING-GUN HARPOON.

Head, common toggle, mortised, pivoted to end of shank. Shank, composed of two conjoined pieces of iron. Forward end, rod

PIERCE'S DARTING-GUN HARPOON—Continued.

iron; rear portion piping, screwed into end of forward part. Iron-link, twisted, attached to middle of shank for iron-strap. Iron-strap, whale-line. Button, felt. Point of toggle fractured. Length, 24½ inches. Provincetown, Massachusetts, 1882. 56227. Gift of I. A. Small. Marked "B N——" (initials of vessel). Formerly used with the first darting gun.

PRUSSIC ACID HARPOONS.

PRUSSIC ACID HARPOON WITH FIXED HEAD.

Consists of a fixed head of highly-tempered steel, with two rigid steel blades slotted on either side at right angles, and two movable barbs pivoted, one on either side, to its rear; and of a sliding attachment, moving independently of the fixed head, composed of two rigid and two movable barbs. The shank is wrought iron, with a socket at one end, and a slot, 8 inches in length, cut by hand, in the other. The piece composing the adjustable barbs is cast with a longitudinal slot on either side, to correspond in size to the sides of the slotted end of the shank. In process of manufacture the sides of the slotted end of the shank are heated and "spread"; the movable barbs having been inserted, the sides are closed permanently to prevent the barbs from being released, but affording ample room for them to move back and forth longitudinally. The vials of poison may be inserted in the space between the fixed head and the independent barbs. When the harpoon is thrown into the whale the action of the flesh and the resistance upon the line draws the adjustable barbs in the direction of the fixed head, crushes the vials, and destroys the whale. Length, 45 inches. Nantucket, Massachusetts, 1882. 56260. Gift of Mr. Joseph B. Macy. Rare. Not used at present.

PRUSSIC ACID HARPOON WITH ADJUSTABLE HEAD.

The head consists of a diamond-pointed, highly tempered piece of steel, with two rigid fin-like blades, or barbs, slotted through it at right angles; two movable flukes, or toggles, pivoted to its rear; and a neck, cast with the head, 9 inches long. The shank terminates in a socket for the pole, and is slotted at the forward end, forming a recess for the neck. The neck is adjusted longitudinally in the slot in such a manner that while it may be moved back and forth with great facility, being held by a metal pin, it cannot be separated from the shank. The neck is fluted or grooved on the two sides which are abreast the openings of the slot near the end of shank. Two vials of poison may be placed respectively in the recesses when the head is in repose. When the instrument is imbedded in the

PRUSSIC ACID HARPOON WITH ADJUSTABLE HEAD—Continued.
flesh of the whale, the counter-resistance upon the line—the whale moving in the direction of the head, and the weight of the boat being brought to bear upon the shank—the adjustable head and neck are drawn out until stopped by the metal pin crushing the vials and destroying the whale. Length, 49 inches. Nantucket, Massachusetts, 1882. 56261. Gift of Joseph B. Macy. Rare. Not used at present.

HARPOONS PROJECTED FROM GUNS.

NON-EXPLOSIVE.

SMITH'S GUN-HARPOON.
Head, or toggle, slotted. Double shank, wrought iron. Endless wire loop for making fast one end of whale-line; leather pad. Provincetown, Massachusetts, 1876. 29396. Gift of Lemuel Cook 2d. Seldom used at present.

ALLEN'S GUN-HARPOON.
Galvanized. Patented December 5, 1848, by Oliver Allen, of Norwich, Connecticut. Fixed head with four barbs. The shank about half its length is composed of an iron rod, which terminates in a head or button and socket; the rear portion of the shank malleable cast iron. At the junction of the two pieces of shank is a collar intended to be used as a *stop* for the *iron-strap*. Not in use at present. Length, 43½ inches. Provincetown, Massachusetts, 1876. 29500. Gift of E. K. Cook.

GUN-HARPOON.
Fixed head, double barbed, diamond point, steel. Two movable flukes, wrought iron, pivoted below head. Shank, double, wrought iron; leather button, screwed. Endless wire loop with leather pad. Length, 16½ inches. Middletown, Connecticut, 1882. 54333. Gift of A. R. Crittenden.

SMITH'S GUN-HARPOON.
Movable barb. Head of shank cast with slot for barb; shank, malleable iron, double, provided with wire loop into which may be bent one end of the iron-strap. Formerly used in connection with the shoulder-gun. Middletown, Connecticut, 1882. 54334. Gift of A. R. Crittenden. Not used at present.

KELLY'S GUN-HARPOON.
Head, toggle, malleable cast iron, mortised. Shank, wrought iron, with collar at butt. Adjustable slide, to which is attached rope-beckets. Button, rubber. Cushion-spring broken. Marked "B. G. L. A. 111." (Initials of brig and number of boat).

KELLY'S GUN-HARPOON—Continued.

Iron-strap consists of two rope loops, or beckets, about 10 inches long, with leather in the eyes where they are made fast to the slide. Length, 26¾ inches. New Bedford, Massachusetts, 1882. 56210. Gift of Thomas Knowles & Co.

GUN-HARPOON WITH FLAT SHANK.

Head, common toggle. Eye made in a piece of strap-iron riveted to the shank, with hole for splicing in iron-strap. Button, leather, attached with screw. Length, 23 inches. New Bedford, Massachusetts, 1882. 56211. U. S. Fish Commission. Designed and manufactured by Captain Eben Pierce, as an experiment. None in use. Point broken off. Has been fired into a gravel-bed.

SMITH'S GUN-HARPOON.

Temple gig, pivoted in the cheeks of the forward end of shank. Shank, double, cast. Wire loop and thimble for iron-strap. Length, 29 inches. New London, Connecticut, 1882. 56212. Gift of Lawrence & Co. Not used at present.

BROWN'S GUN-HARPOON.

Fixed head, diamond point, with projecting lanceolate blades on sides and top; two toggles in rear pivoted to a fixed head. Shank, flat; fluted on both sides; two holes near the head and one in rear end for iron-strap. Length, 35½ inches. Fairhaven, Massachusetts, 1882. 56213. Gift of Luther Cole.

BROWN'S GUN-HARPOON.

Head, double-barbed, diamond pointed, surmounted at right angles by lance-like cutting-point. Head cast in one piece. Two movable barbs, pivoted, one on either side, in the rear of fixed head. Shank, cast-iron, flat and fluted; two holes at forward end of shaft and one hole at the butt for making fast the iron-strap. Button, iron. Length, 37 inches. New Bedford, Massachusetts, 1882. 56214. Gift of John McCullough.

SHOULDER GUN-HARPOON.

Head, common toggle; malleable cast-iron; mortised and pivoted to shank. Shank, single, round, wrought iron. Button, iron. Marked "B. G. A." (Initials of name of brig), "111" (No. of boat), "Dean & Sawyer" (Manufacturers). Length, 37 inches. New London, Connecticut, 1882. 56215. Gift of C. A. Williams & Co.

BROWN'S SHOULDER GUN-IRON.

Head fixed, four barbs, the main portion of which consists of the two ordinary harpoon barbs surmounted at right angles by

BROWN'S SHOULDER GUN-IRON—Continued.
> two other smaller barbs which are brazed to the former at the extreme end. Barbs, steel; head welded to shank. Shank, forged wrought iron; fluted on both sides for the reception of the iron-strap which accompanies the harpoon when placed in the gun; rear extremity of the shaft terminates in a button or boss; two holes near the head of the shaft, and one hole in the rear end, to which is made fast the iron-strap. Length, 37¼ inches. New Bedford, Massachusetts, 1882. 56216. Gift of John McCullough. Seldom used at present.

SHOULDER GUN-HARPOON.
> Head consists of two barbs, forward extremity diamond pointed, rear extremity forming a fluke, the whole acting upon the principle of a toggle; slotted and pivoted to shank; shank, double, cast-iron. Length, 36 inches. Nantucket, Massachusetts, 1882. 56218. Gift of Joseph B. Macy.

BROWN'S GUN-HARPOON.
> Used in connection with the shoulder-gun. Head, double-barbed, fixed, diamond point; two toggles, one on either side, pivoted to rear portion of main head, and slotted to fold over the plates. Shank, cast-iron, fluted on both sides; two holes in the head of shaft, and one hole near the butt into which is spliced the iron-strap. Shank has been broken and brazed, as the quality of iron would not permit welding. Length, 31¾ inches. New London, Connecticut, 1882. 56219. Gift of Messrs. Lawrence & Co.

SMITH'S GUN-HARPOON.
> Head malleable iron, three cutting-edges; flukes, malleable iron, pivoted in slot in forward part of shank. Shank, double; cast-iron. Wire loop for attaching the tow-line. Length, 25½ inches. Provincetown, Massachusetts, 1882. 56220. U. S. Fish Commission. Seldom used at present.

THE MASON GUN-HARPOON.
> Cast-iron. Lance-like cutting point and two movable barbs. Shank, grooved; eye in butt for rope. Length, 22 inches. New Bedford, Massachusetts, 1882. 56221. Gift of H. W. Mason. Recent invention.

KELLY'S GUN-HARPOON.
> Head, common toggle, malleable cast-iron, mortised; shank, wrought-iron, with collar at butt; adjustable slide, with rigid eyes in which loops, or beckets, made of rope are fastened; spiral cushion-spring to neutralize the shock of sudden stoppage; button wanting. Length, 26½ inches. New Bedford, Massa-

KELLY'S GUN-HARPOON—Continued.

chusetts, 1882. 56223. Gift of Thomas Knowles & Co. Patented by Zeno Kelly, New Bedford, Massachusetts, December 3, 1867. Seldom used at present.

SWIVEL GUN-HARPOON.

Head, common toggle, slotted and pivoted to shank. Shank composed of a single rod of wrought-iron, with forged iron button at rear extremity. Marked "B. G. A., 11." "Dean & Sawyer" (Manufacturers). Length, 48 inches. New London, Connecticut, 1882. 56224. Gift of C. A. Williams & Co.

SWIVEL GUN-HARPOON.

Head consists of toggle, double-barbed; toggle slotted and pivoted to end of shank; shank double, wrought-iron, welded at both ends; button at rear end. Length, 46 inches. Edgartown, Massachusetts, 1882. 56225. Gift of Messrs. Osborn & Co.

SMITH'S GUN-HARPOON.

Malleable iron; one pivoted malleable cast-iron fluke. Shank, double; head with slot for fluke. Wire loop for iron-strap, wrapped with marline. Length, 25½ inches. Provincetown, Massachusetts, 1882. 56228. Gift of Stephen Cook. Seldom used at present.

ALLEN'S GUN-HARPOON.

Head, fixed; four barbs. Forward part of shank, wrought iron; rear part, iron piping, welded to rear end of forward part. Length, 44¼ inches. Provincetown, Massachusetts, 1882. 56229. U. S. Fish Commission. None in use at present. Erroneously called Broomstick-lance.

ALLEN'S GUN-HARPOON.

Head, fixed; four barbs. Shank consists of two pieces of conjoined iron; for about one-half its length it is composed of an iron rod, with a head or button at its rear end, which is intended as a *stop* for the iron-strap; the rear part of the shank is cast with four strips of metal meeting each other at right angles; the forward end fitting in the socket of the first piece. Length, 44¾ inches. New Bedford, Massachusetts, 1882. 56230. U. S. Fish Commission. Not used at present. Formerly fired from shoulder-guns.

SMITH'S GUN-HARPOON.

Head, slotted; shank, double, wrought iron. Loop for making fast end of tow-line, wire, wrapped with spun-yarn. Button, wanting. Length, 26¼ inches. Provincetown, Massachusetts, 1882. 56232. U. S. Fish Commission. Seldom used at present.

BROWN'S GUN-HARPOON.
  With iron-strap attached. Head, diamond point, fixed; toggle pivoted on either side. Shank, flat, fluted on both sides; one hole at butt, and two holes in forward end into which is spliced the iron-strap. Iron-strap, 1½ inch rope, free end unlaid. Length, 35½ inches. New London, Connecticut, 1882. 56234. Gift of Lawrence & Co. Seldom used at present.

GUN-HARPOON.
  Temple gig; shank, single, round; button, iron. Hole in end of shank for rope. Length, 42½ inches. New Bedford, Massachusetts, 1882. 56235. Gift of Thomas Knowles & Co.

SWIVEL GUN-HARPOON.
  Head, wrought iron, double-barbed. Shank, malleable iron, cast; double or slotted. Loop with two eyes, wire, wrapped with wire; iron thimble attached, with rope for making fast the whale-line. Marked "S. Lydia." Length, 48 inches. Edgartown, Massachusetts, 1882. 56236. U. S. Fish Commission. Cut from a dead whale.

PIERCE'S GUN-HARPOON.
  Head, or toggle, malleable cast iron. Shank, forward part wrought iron; rear part, piping. Felt tip. Iron-strap, whale-line, spliced at one end in eye in shank, and with an eye-splice in the other for making fast one end of the tow-line. Length, 26 inches. New Bedford, Massachusetts, 1882. 56238. U. S. Fish Commission. Manufactured by Captain Eben Pierce. Old style, formerly used with the first darting gun.

ALLEN'S GUN HARPOON.
  Four fixed barbs. Shank in two sections; forward part wrought-iron terminating in a socket, into which is fitted the rear part; provided with a fixed iron collar to be used as a stop for the iron-strap. Rear portion of shank wood with iron ferule. Length, 46¼ inches. New Bedford, Massachusetts, 1882. 56239. U. S. Fish Commission. Formerly used with the shoulder-gun.

SWIVEL GUN-HARPOON.
  Fixed head, double barbed; double-shank; branded "Tinkham" (Manufacturer). Head, wrought-iron; shank, malleable cast iron. Length, 48½ inches. New London, Connecticut, 1882. 56240. Gift of Messrs. Lawrence & Co.

SWIVEL GUN-HARPOON.
  Double shank. Shank and toggle malleable cast iron. Marked "B. K. M. T." (Initials of bark to which the harpoon belonged). Length, 51 inches. Nantucket, Massachusetts, 1882. 56241. Gift of Joseph B. Macy.

THE CALIFORNIA GUN-HARPOON.

Head, common toggle, with flaring barb at rear, wrought iron, mortised; shank, double; wire loop, served with twine; iron-strap and eye-splice. Length, 50 inches. Length of toggle, $9\frac{3}{8}$ inches. San Miguel, California, 1882. 72753. Manufactured and presented by George W. Proctor. Used in connection with the swivel gun at the whaling stations on the coast of California, and other points on the northwest coast, for the capture of the California grays, humpbacks, finbacks, and right whales.

EXPLOSIVE HARPOONS.

THRUST BY HAND.

BARKER'S BOMB-HARPOON.

A harpoon with an adjustable explosive lance-like head, intended to be used for simultaneously "fastening to" and killing whales. Length, 33 inches. Provincetown, Massachusetts, 1882. 56370. Gift of D. Connell. This kind of instrument is intended to be darted by hand. The resistance of the skin of the whale upon the brass cross-piece draws back the discharging-wire which extends parallel to the shank, causing the hammer to strike upon the cap, and explodes the bomb, or head, in the whale. When the whale runs, or sounds, the traction of the line upon the harpoon expands the brass barb, or fluke, and prevents the instrument from being withdrawn. Reloaded by the substitution of a new head. Patented by Silas Barker, February 21, 1865.

FREEMAN'S BOMB-HARPOON.

An instrument with an explosive head for killing whales. Consists of a chambered head, or magazine, which, when loaded, contains a charge fully equal to three-fourths of a pound of powder; a shank with a tubular head and two small rigid barbs, and socket for the pole. The inside mechanism consists of a time-fuse, which extends from the shank into the magazine, a nipple for the percussion-cap, an actuating spring, and other appliances for releasing the cock, which are concealed in the recessed head of the shank. The trigger, or lever-fluke, is fastened by a hinge-pin immediately in rear of the lance-bomb. The action of the flesh as the instrument enters the whale presses down the trigger, or fluke, in a line with the shank, and automatically explodes and impels the head. Reloaded by substituting new heads. Length, $40\frac{1}{2}$ inches. Brewster, Massachusetts, 1876. 42762. Made by Freeman & Lincoln. This kind has been employed for killing the finback whale on the coast of Cape Cod, Massachusetts. Patented by Charles Freeman, Brewster, Massachusetts, May 7, 1872.

PROJECTED FROM GUNS.

MASON'S HARPOON.

Designed for an improved swivel-gun. Consists of a point with three cutting edges and cast-iron bomb, cast-iron shank with four parallel grooves on the sides, and an eye at the butt for the iron-strap. Two movable flukes are fastened with a set-screw to the forward end of the shank in rear of the bomb. Length, 31¾ inches. New Bedford, Massachusetts, 1882. 56376. Manufactured and presented by H. W. Mason. Recent invention.

WALRUS HARPOONS.

Common toggle-irons; shanks, wrought iron; toggles, malleable cast iron, mortised; stamped "L. Cole" (Manufacturer). Total length (56415), 22 inches; total length (56413), 22½ inches; length of toggles, 6 inches. Fairhaven, Massachusetts, 1883. U. S. Fish Commission. Used by whalemen in the Arctic regions in the capture of walrus.

IRON SHEATHS.

SHEATH FOR HARPOON.

Made for the harpoons with fixed heads. Wood, covered with canvas, painted black. Length, 7 inches. 56402. U. S. Fish Commission.

SHEATH FOR TOGGLE-IRONS.

A kind of sheath for the toggle-iron, usually made at sea, to prevent the heads of the harpoons from rusting, and from inflicting flesh-wounds upon the men when handling them. Common to all whaling vessels. Length, 10 inches. New Bedford, Massachusetts, 1882. 56400 and 56401. Gift of I. H. Bartlett & Sons.

WHALING-GUNS.

ORDNANCE.

SWIVEL-GUNS.

Barrel, stub-twist; stock, hard wood; guide for taking aim, brass, extending along and screwed to barrel; elevated at rear end. Barrel fastened to stock by bolts and lugs. Breech-plug chambered and screwed into the barrel. Two nipples. Flash-pan, brass, hinged to rear of elevated sight. Barrel stamped "W. Greener, maker, Birmingham, 1853." Length, 51 inches; weight, 56 pounds. New Bedford, Massachusetts, 1882. 56342; ramrod, 56343; swivel, 56344; gun-strap, 56345; wrench, 56346. U. S. Fish Commission. Mounted on a swivel which is inserted in a loggerhead made fast to the clumsy-cleat in the bow of the boat, and used in discharging gun-harpoons and ex-

FISHERIES OF THE UNITED STATES. [51]

SWIVEL-GUNS—Continued.
plosive lances for fastening to and killing whales. When cocked, the hammer is held stationary by a small iron pin inserted in a hole in the stock, as a precaution against premature discharge. The gunner having taken aim at a whale, at the opportune moment removes the pin and fires the gun by means of a laniard fastened to a hair-trigger.

SMALL-ARMS.

SHOULDER-GUNS—MUZZLE-LOADING.

BRAND GUN, No. 1.
Barrel, cast-steel; front part of barrel round. Elongated thimble for ramrod. Ramrod, hickory, with brass thimble and screw. Skeleton stock, iron, screwed to barrel. Guard-plate, steel. Rigid eye for laniard. Laniard attached. Patent breech. Stock and barrel "blued." Length, 38 inches; weight, 23 pounds. Norwich, Connecticut, 1883. 56336. Gift of Junius A. Brand. Three drachms of powder ("Sea-shooting, F. F. G.") is recommended as a charge for shooting a bomb-lance.

BRAND GUN, No, 2.
Barrel, cast-steel; front part of barrel round. Ramrod, hickory. Skeleton stock, iron. Guard-plate, steel. Rigid eye for laniard. Patent breech. Stock and barrel "browned." Length, 36 inches; weight, 19½ pounds. Norwich, Connecticut, 1876. 24986. Gift of C. C. Brand. Used with C. C. Brand's improved bomb-lance: 24989—wad-cutter (inside); 24992—prepared wads; 24990, 24991, screw-drivers. Three drachms of powder (Sea-shooting, F. F. G.) is recommended by the manufacturer as a charge for shooting a bomb-lance.

SHOULDER-GUN WITH BRASS STOCK.
Barrel, cast-steel, octagonal. Rear and front sights. Two thimbles for ramrod. Ramrod, hickory, with brass thimble and double worm-screw. Underside of barrel grooved, for ramrod. Stock, gun-metal, cast with breech-plug and rigid eye for laniard. Grip wrapped with marline. Lock, common percussion. Length, 35½ inches; weight, 28 pounds. New Bedford, Massachusetts, 1882. 56335. Gift of William Phillips & Son. These guns were formerly used, but very little is known of their history.

SHOULDER-GUN WITH WOODEN STOCK.
Barrel, smooth-bored, cast steel. Hind and fore sights. Patent breech. Stock, walnut. Mountings, steel and brass. Rigid eye on underside of stock for laniard. Percussion lock. Guard-plate, steel. Length, 38½ inches; weight, 25 pounds. New

SHOULDER-GUN WITH WOODEN STOCK—Continued.
> Bedford, Massachusetts, 1882. 56340. U. S. Fish Commission. Not used at present; formerly employed in connection with gun-harpoons and bomb-lances.

BROWN'S WHALING-GUN.
> Stock, barrel, and guard-plate, gun-metal; trigger-guard fastened to stock with three screws; rigid eye on trigger-guard for laniard; front and rear sights; breech-plug cast with stock; stock recessed for two nipples; stock and barrel connected by a screw-joint; muzzle reinforced with a gun-metal band. Stamped "Robert Brown, New London, Connecticut." Length, 46 inches; weight, 36 pounds. New Bedford, Massachusetts, 1882. 56341. U. S. Fish Commission. A kind of gun formerly used for projecting gun-harpoons. Charged with ordinary powder and discharged from the shoulder. Not used at present.

BROWN'S GUN (short).
> Stock, barrel, and trigger-guard, gun-metal. Stock screwed to barrel by means of permanent breech-plug cast with the stock. Small eye cast with trigger-guard for laniard. Laniard, lance-warp. Copper band at muzzle. Length, 36¼ inches; weight, 33 pounds. New London, Connecticut, 1882. 56339. Gift of Messrs. Lawrence & Co. Stamped "C. A." and supposed, from its resemblance in almost every detail to the Brown gun, with the exception of the length of the barrel, that it may have been manufactured by Robert Brown, New London, Connecticut. Not used at present.

RIFLED WHALING-GUN.
> Barrel, cast steel, nine grooves; stock, walnut; rigid eye for laniard. Length, 38 inches; weight, 18 pounds. New Bedford, Massachusetts, 1882. 56338. Gift of Edward P. Haskell, jr. Manufactured by Grudchos & Eggers, New Bedford, Massachusetts. This kind of gun was formerly used, to a limited extent, however, by American whalemen, but has been supplanted by the more recently improved patterns, the principal objection being the rifled barrel. Aside from the trouble and delay occasioned by a foul barrel, it was found impossible to use any other lances than those made especially for it. Some of these rifles have been transformed into the class of guns known as "smooth-bores" by removing the grooves; but they were not regarded with much favor, and the manufacture of rifled guns for whaling has virtually ceased. Kentucky rifle powder, F. F. G., was used, the amount for each load being graduated by the "charger" of the flask furnished with the gun.

FISHERIES OF THE UNITED STATES.

BREECH-LOADING GUNS.

PIERCE AND EGGERS' GUN.

One of the latest improved shoulder-guns, used in connection with the Pierce and Brand explosive lances. Skeleton stock. Stock, barrel, breech-block, and trigger, gun-metal; barrel reinforced. The gun is loaded by inserting a cartridge—Winchester No. 8, central fire—in the breech, and the lance in the muzzle, and discharged as an ordinary shot-gun, the cartridge being ignited by a firing-pin striking a percussion-cap. Length, 36½ inches. Weight, 24 pounds. New Bedford, Massachusetts, 1882. 56337; cartridge-holder, 56347; cartridge-primer, 56348; cartridge 56349; charger, 56350; wads, 56351; gun-stick, 56344. Deposited in part by the U. S. Fish Commission and S. Eggers, Sr. Patented February 12, 1878, by Eben Pierce and S. Eggers, Sr. Manufactured by S. Eggers, New Bedford, Massachusetts. The gun in its present condition is the same as when used by whalemen.

CUNNINGHAM AND COGAN'S GUN.

Skeleton stock, cast iron, painted black; stock and breech-piece cast in one piece, with a small rigid eye at rear of guard-plate for laniard; barrel, steel, reinforced and screwed to the stock; breech-block, containing firing-pin, hinged to stock, and, when closed, held by a snap-spring; central-fire cartridge. Length, 33 inches. Weight, 27 pounds. New Bedford, Massachusetts, 1882. 56334. Gift of William Lewis. Patented by H. W. Chapman, Newark, New Jersey, May 15, 1877, and manufactured by Patrick Cunningham, New Bedford, Massachusetts. Used principally by the crews of the steam barks in the Arctic regions; discharges Cunningham & Cogan's explosive lance. This is the form commonly in use, and was selected from a lot about to be sent to a whaling-vessel.

DARTING-GUNS.

PIERCE'S DARTING-GUN (old).

Breech, brass, cast with breech-piece. Barrel, steel, screwed to breech-piece. Rear end of the gun terminates in a conical socket, into which may be fitted the pole or handle. A vertical slot is cut through the breech for the reception of the hammer, which was pivoted and retained in its firing position by the rod or trigger. Hammer, wanting. Trigger projects over the muzzle, and moves freely back and forth in a guide near the end of the barrel. A sleeve of metal, or other suitable material, was intended to fit over the breech, or lock-case, to render it water-tight. The harpoon is of the pattern known as the "Temple-gig." Toggle, malleable cast iron, pivoted in

PIERCE'S DARTING-GUN (old)—Continued.
the cheeks of the forward end of the shank. Shank composed of two pieces of conjoined iron: First half, wrought iron, slotted near its rear end for the iron arm with rigid eye to which the iron-strap should be made fast, and provided with a female screw in a recess in the rear end. Rear part of shank cast, and screwed to the forward half of the shank. Length of gun, 17½ inches; length of trigger, 27½ inches; length of gun-harpoon, 23¼ inches. Provincetown, Massachusetts, 1882. Gun, 56331; harpoon, 56441. Presented by Mr. Seth Smith. The gun having been charged with ordinary powder and the iron inserted, it is darted and automatically discharged by the long wire trigger coming in contact with the whale. An imperfect model of the first darting-gun with gun-harpoon, invented by Capt. Eben Pierce, August 22, 1865. Not used at present.

PIERCE'S MUZZLE-LOADING DARTING-GUN.
Old style, with pole, harpoon, and darting-bomb; muzzle-loading. Gun: barrel, lock-case, socket for pole, socket pieces or lugs for harpoon, and one forward guide for trigger, gun-metal; two after-guides for trigger, brass; bottom of lock-case, brass, soldered; firing-pin and trunnion, brass; lock-case (cover for excluding moisture), leather with brass catch. Hammer, concealed in lock-case, gun-metal. Trigger, projecting beyond muzzle, steel rod. Harpoon, common toggle-iron; rear end made to fit socket-pieces on the muzzle of the gun, and provided with a projecting eye, in which the iron-strap is made fast. Toggle branded "Macy" (Manufacturer). Length of gun, 19¾ inches; length of pole, 56¼ inches; length of trigger, 34 inches; length of harpoon, 30 inches; length of darting-bomb, 15½ inches. New Bedford, Massachusetts, 1876. 25252. U. S. Fish Commission. Patented and manufactured by Capt. Eben Pierce, New Bedford, Massachusetts. Stock number 476.

CUNNINGHAM'S DARTING-GUN.
Breech-loading hinge gun, with harpoon, strap, and bomb-lance. Gun: barrel, socket, breech-snap, hinge, and lugs, gun-metal; trigger, steel rod, projecting beyond the muzzle. Lance and cartridge combined. Harpoon, common toggle-iron; two barbs on the toggle; mortised head; rear end of shank made to fit the lugs of the gun. Eye for rope-strap. Toggle branded "J. A. S." (John A. Sawyer, manufacturer). Iron-strap, whale-line; one end of strap bent into the eye of harpoon and the other provided with an eye-splice into which one end of the whale-line is intended to be fastened. Length of gun, 15½

CUNNINGHAM'S DARTING-GUN—Continued.

inches; length of trigger, 21 inches; length of harpoon, 34 inches; length of strap, 64 inches. New Bedford, Massachusetts, 1882. Gun, 56332; harpoon, 56352; lance, 56333; iron-strap, 56353. U. S. Fish Commission. Patented (1882) and manufactured by Patrick Cunningham, New Bedford, Massachusetts. Stock number of gun, 107.

RECHTEN'S DARTING-GUN.

Gun, and guides for trigger, cast in one piece, gun-metal. Trigger, or sliding-rod, cast-steel, with a spiral brass cushion-spring. The lock-portions are concealed within the stock and protected from exposure by a hinged cover, which, when closed, fastens with a snap-spring. Stamped on breech "Patd. Dec. 7th, 1869;" stamped on barrel, monogram "J. P. R.," and stamped on sliding-rod, or trigger, " F. Hobson; warranted Sheffield cast-steel." Harpoon consists of a double-barbed fixed head; a short neck with an eye in its center, to which is welded a loose link with an eye in the rear for the iron-strap. The gun may be thrust by hand and the harpoon automatically discharged into the whale by impact. Length of gun, 9¾ inches; length of harpoon, including link, 11¼ inches. New Bedford, Massachusetts, 1882. 56330. U. S. Fish Commission. An imperfect model of a kind of darting-gun patented by John P. Rechten, of New York, N. Y., December 7, 1869. Not in use.

BURSTED BARREL OF A WHALEMAN'S DARTING-GUN.

Barrel of a darting-gun fractured by premature explosion of a bomb-lance when darted at a whale. Brought home by a whaleman as an interesting "curio." Portion of hickory pole in the socket. Length, 10 inches. New Bedford, Massachusetts, 1882. 56328. Gift of Jonathan Bourne.

MODEL OF DARTING GUN.

Said to be the original model of the first breech-loading darting-gun. Wood, in two sections. Hammer, wood, operated by a small piece of wire and a steel spring. Length, 16 inches. New Bedford, Massachusetts, 1882. 56329. Gift of Patrick Cunningham. Made by a whaleman on board ship in the Arctic regions. Scrimshaw work.

ROCKET GUNS.

ROCKET GUN.

A stockless gun with a barrel of such shape and proportion as to balance on the shoulder of the gunner; designed to throw large rockets and shells; barrel, sheet-copper, cylindrical; two rods project behind the barrel and are fastened to an iron plate; barrel encircled with two wide transverse flanges, the lower one

ROCKET-GUN—Continued.

fixed, and the upper one hinged in such a manner that when the gunner is taking aim it lies parallel to the barrel, but is thrown up vertically by the action of the rocket to protect the gunner from the "backfire" of the rocket. The gun is discharged by firing the pistol through a hole made in the stock into the rocket. U. S. Fish Commission. 56327. Patented January 22, 1861, by Thomas W. Roys, of Southampton, New York. Employed principally in whaling off the Northwest coast. Successfully used from the decks of steamers, for which it was designed.

WHALEMAN'S LANCES.

THRUST BY HAND.

EXPLOSIVE.

KELLEHER'S HAND BOMB-LANCE.

Consists of a lance-head, a tubular magazine, and the ordinary harpoon shank, secured to a white-ash handle. A sliding clamp attached to a wire by impact explodes the bomb by means of a common friction-primer, such as is used for discharging pieces of artillery. Socket served with marline to prevent iron-rust. Lance-strap spliced around the socket, seized to the handle in three places, and projecting through a hole at the butt. Length of lance and shank, 48¼ inches; length of pole, 70 inches. New Bedford, Massachusetts, 1882. 56359. Gift of Daniel Kelleher. (Patented by Daniel Kelleher March 26, 1878.) This instrument may be used as an explosive or non-explosive hand-lance. As there is no "backward kick," or recoil, the operator grasps the handle when the lance explodes.

NON-EXPLOSIVE.

HAND-LANCE.

A kind of lance with a short wide blade, formerly used for killing whales. Superseded by the explosive lance. Length, 68 inches. New Bedford, Massachusetts, 1876. 25611. Presented by W. H. Cook & Co. Primitive model used by the New Bedford whalemen.

HAND-LANCE.

A nickel-plated hand-lance used in giving the death-wound. Length, 5 feet 9 inches. Fairhaven, Massachusetts, 1882. 56357. Manufactured and presented by Luther Cole.

HAND-LANCE.

A kind of lance formerly relied upon altogether for killing whales. Length, 5 feet 3 inches. New Bedford, Massachusetts. 56389. Gift of James Barton. Manipulated by the officer of the boat.

FISHERIES OF THE UNITED STATES.

HAND-LANCE.

Head, cast steel; shank and socket, wrought iron. Handle, wood. New. Total length, 13 feet. Fairhaven, Massachusetts, 1883. 56418. U. S. Fish Commission. Formerly extensively used for killing the whale. Stamped "Luther Cole" (Manufacturer.)

HAND-LANCE AND POLE.

Head, cast steel; shank and socket, wrought iron. Handle, wood. New. Total length, 13 feet. Fairhaven, Massachusetts, 1883. 56418. U. S. Fish Commission. Formerly extensively used for killing the whale. Stamped "Luther Cole" (Manufacturer.) Rigged for use.

SEAL, SEA-ELEPHANT, AND WALRUS LANCES.

THRUST BY HAND.

SEAL-LANCE.

Long head; diamond point; common shank and socket. Manufactured by James Barton for the New London sealers. New. Length, 32½ inches. New Bedford, Massachusetts, 1882. 56366. U. S. Fish Commission.

SEAL-LANCE.

A kind of lance with a short shank which may be used in killing seal, sea elephant, or walrus. Socket with an extended sleeve. Length, 28½ inches. New Bedford, Massachusetts, 1883. 56367. Gift of Luom Snow & Son. Old. Has been used.

SEAL-LANCE.

A kind of lance which may be used for killing seal, sea elephant, or walrus. Spoon-shaped head, and extended sleeve on socket. Used by New Bedford sealers. Length, 24 inches. New Bedford, Massachusetts, 1883. 56368. Gift of Luom Snow & Son. Old. Obtained from a New Bedford sealing vessel.

SEAL-LANCE.

A kind of lance which may be used for killing seal, sea elephant, or walrus. Socket with extended sleeve. Section of pole attached. Length, 24 inches. New London, Connecticut, 1882. 56369. Gift of Lawrence & Co. Old. Has been used. Obtained from a New London sealer.

LANCES FOR STOPPING RUNNING WHALES.

FLUKE-LANCE.

A kind of lance intended to take the place of the thick boat-spade. Manufactured to gratify a whim of a whaleman. None in use. Length, 5 feet. New Bedford, Massachusetts, 1882. 56358. Gift of James D. Driggs, manufacturer.

EXPLOSIVE LANCES.

PROJECTED FROM GUNS.

ALLEN'S BOMB-LANCE.

An example of the first patented bomb-lance used by American whalemen, for which letters patent were granted Oliver Allen, Norwich, Connecticut, September 19, 1846. (U. S. Patent Office, No. 4764.) Rare. Familiarly known to whalemen as the "Broomstick lance." Length, 42 inches. Nantucket, Massachusetts, 1882. 56372. Gift of Joseph B. Macy. The instrument consists of a steel lance-head joined by a closely-fitting tenon to a cylindrical bomb which is composed of a piece of iron piping; a neck of smaller diameter than the bomb, brazed to rear of bomb, and a long tubular shank, wood, confined to the rear end of neck with a brass ferule. The rear end of the shank terminates in a small metal button through the axis of which a small aperture is made. The time-fuse, inclosed in the tubular shank, is ignited by the flash of the discharge of the gun, which passes through the vent hole of the metal button.

BRAND'S BOMB-LANCE NO. 1 (OLD MODEL).

Primarily intended to be used with Brand's No. 1 gun. Patented in 1852 by C. C. Brand. Length, 16¾ inches. Norwich, Connecticut, 1883. 56388. Manufactured and presented by Junius A. Brand.

BRAND'S BOMB-LANCE NO. 1 (SHORT).

May be used with Brand's No. 1 Muzzle-loading gun, Pierce & Eggers' Breech-loader, or Pierce's Darting-gun. New model. Patented in 1879. Length, 16¾ inches. Norwich, Connecticut, 1883. 56387. Patented, manufactured, and presented by Junius A. Brand.

BRAND'S BOMB-LANCE NO. 1 (LONG).

May be used with Brand's No. 1 Shoulder-gun, the Pierce & Eggers' Breech-loading gun, or the Pierce Darting-gun. New model. Patented in 1879. Length, 18¾ inches. Norwich, Connecticut, 1883. 56386. Patented, manufactured, and presented by Junius A. Brand.

BRAND'S BOMB-LANCE NO. 2.

Used with Brand's No. 2 Shoulder-gun. Old model. Patented in 1852 by C. C. Brand. Length, 21½ inches. Norwich, Connecticut, 1883. 56384. Manufactured and presented by Junius A. Brand.

FISHERIES OF THE UNITED STATES.

BRAND'S BOMB-LANCE No. 2.
Used with Brand's No. 2 Shoulder-gun. New model. Patented in 1879. Length, 21¼ inches. Norwich, Connecticut, 1883. 56385. Patented, manufactured, and presented by Junius A. Brand.

BRAND'S BOMB-LANCE No. 3.
Used with Brand's No. 2 Shoulder-gun. Old model. Patented in 1852 by C. C. Brand. Length, 24 inches. Norwich, Connecticut, 1883. 56382. Manufactured and presented by Junius A. Brand.

BRAND'S BOMB-LANCE No. 3.
Used in connection with Brand's No. 3 Shoulder-gun. New model. Patented in 1879. Not much used at present. Length, 24 inches. Norwich, Connecticut, 1883. 56383. Patented, manufactured, and presented by Junius A. Brand.

BRAND'S BOMB-LANCE No. 4.
Used in connection with Greener's swivel-gun. New model. Patented in 1879. Length, 34 inches. Norwich, Connecticut, 1883. 56381. Patented, manufactured, and presented by Junius A. Brand.

PIERCE'S BOMB-LANCE.
Main portion, or powder chamber, brass tubing; anterior end provided with nipple for percussion-cap and time-fuse. Rear end or tail-piece, composition metal; fluted sides with longitudinal slots for reception of the wings. Guide-wings, sheet-brass, fastened to brass wires; closed by a brass ring when placed in the gun, and expand radially from a common center when discharged. Lance-point, composition metal; four cutting edges; recessed, containing a hammer secured by a wooden pin, which is broken by the concussion of the explosion of the charge, and explodes the cap on the nipple in the end of the shank, communicating the fire to the magazine by means of the time-fuse. Button, sole leather, fastened with a screw. Length, 19 inches. New Bedford, Massachusetts, 1882. 56355. Manufactured and presented by Captain Eben Pierce. Used with Pierce & Eggers' Shoulder-gun. Made in three sections. The lance is loaded by detaching the rear section, and capped by detaching the cutting-point.

CUNNINGHAM & COGAN'S BOMB-LANCE.
An improved bomb (with rubber feathers) and cartridge combined, used in connection with Cunningham & Cogan's breech-loading gun, patented December 28, 1875. Length, 16¼ inches. New Bedford, Massachusetts, 1882. 56371. Gift of William Lewis. (Patented and manufactured by Patrick Cunningham).

GRUDCHOS & EGGER'S BOMB-LANCE.

A kind of lance invented and manufactured by Julius Grudchos & Selmar Eggers, of New Bedford, Massachusetts, to be used in connection with the rifled gun. It was an experiment, ended in failure, and has been abandoned. Length, 15¼ inches. New Bedford, Massachusetts, 1882. 56378–9. Gift of Frederick S. Allen. Patented May 26, 1857. Head or blade, steel; lanceolate. Shank forming the magazine, iron tubing. Rear section contains the mechanical contrivances for exploding the bomb. Extreme rear end terminates in a butt, technically termed a "ball," made of lead, with spiral elevations to fit the grooves of the rifle, thereby giving the lance, when discharged from the gun, a rotary motion to prevent it striking sidewise. A groove is made around the ball for the reception of oiled yarn, which is intended as a wadding, as well as for cleaning the rifle. Trigger pivoted at one end in a slot near the extremity of the instrument. The trigger remains in repose, parallel to the shank, when the lance is placed in the barrel; but upon entering the whale it is elevated to an angle of about 40 degrees by the resistance of the flesh and automatically explodes the lance by striking a percussion-cap.

EXPLOSIVE GUN-LANCE.

A kind of explosive lance, the record of which is very little known. Consists of two conjoined parts; the forward half, or magazine, malleable iron, cast with the head, which has four cutting edges; the rear section, or fuse-shaft, cast iron, fluted on three sides for the ropes (which are placed in the gun with the lance), and attached to the bomb with a screw-joint. The rear extremity of the fluted elevations is perforated with three holes through which the strands of rope are rove and braided. Time-fuse inclosed in fluted tubular shank. Cork shoe, or button. Length, 33 inches. Fairhaven, Massachusetts, 1882. 56380. Gift of Luther Cole. As the lance has no barbs, it is evident that the braided rope-tails (which were of a uniform size and so systematically arranged round the axis of the shank) were intended to act in the capacity of wings, by dragging behind the lance with equal force to keep it in a true course during its flight.

EXPLOSIVE LANCES RECOVERED FROM WHALES.

BRAND'S LANCE, No. 1.

Old, not exploded. Cut from a dead whale. Wings burned by the flash of the gun. Length 16 inches. Edgartown, Massachusetts, 1882. 56361. Gift of C. B. Marchant.

BRAND'S LANCE, No. 2.

Old, not exploded. Cut from a dead whale. Rubber wings scorched by the discharge of the gun. Length, 21 inches. Edgartown, Massachusetts, 1882. 56360. Gift of C. B. Marchant.

SERIES OF EXPLODED BOMB-LANCES CUT FROM DEAD WHALES.

Three with rubber wings, and one with metal wings. Fragmentary pieces. New Bedford, Massachusetts, 1882. 56362–'3–'4–'5. Gift of Patrick Cunningham.

NON-EXPLOSIVE LANCES.

PROJECTED FROM GUNS.

GHENN'S LANCE.

Stock, wood, slotted longitudinally the entire length for the reception of a small line, one end of which is made fast to the butt; the other has an eye-splice for bending on the lance-warp. This line, or "lance-strap," having been placed in the longitudinal slot, a strip of paper is pasted over it to hold it in a proper position when loaded in the gun. The rear end of the shank is slotted through the wood to form receptacles for the wings. The wings, two in number, are made of tin soldered to wires, which latter act as springs to compress the wings when placed in the gun-barrel, and to elevate them radially when the instrument is projected. The head, or cutting-point, resembles in shape that of the old double-barbed hand-harpoon, but is smaller; barbs slightly recurved; short neck, terminating in a socket for the reception of the forward end of the wooden shank and strengthened by a lead ferule. Length, 23¼ inches. Provincetown, Massachusetts, 1882. 56356. Gift of Seth Smith. One of the original models. Manufactured and used, to a limited extent, by Captain Josiah Ghenn, of Provincetown, Massachusetts. In about 1849 Captain Ghenn made several of these lances to be fired from a shoulder-gun into the whale after it had been harpooned; but, upon being notified that this pattern was an infringement on C. C. Brand's patent, he discontinued their manufacture.

BROWN'S NON-EXPLOSIVE GUN-LANCE.

Head double-barbed, fixed, with lanceolate blades. Entire head cast iron (case-hardened). Shank, cast iron, flat and fluted on both sides for the reception of rope-tails, which are intended to be used as wings. Eye in the rear end of shaft. Button wanting. Detachable button. Length, 36¼ inches. New Bedford, Massachusetts, 1882. 56217. Gift of Thomas Knowles & Co. Patented by Robert Brown, New London, Connecticut, August 20, 1850. Not used at present.

DARTING-BOMBS.

PIERCE'S DARTING-BOMB.

A kind of explosive lance known as the "darting-bomb," used in connection with the darting-gun for killing whales. Patented and manufactured by Capt. Eben Pierce. Length, 15½ inches. New Bedford, Massachusetts, 1876. 25252. U. S. Fish Commission. Substantially the same as the other kind of the Pierce lance, with the exception that, as it is projected when the muzzle of the gun is in contact with the whale, the wings are considered superfluous, and are not used on this pattern.

BRAND'S DARTING-BOMB.

A kind of explosive lance used in connection with Pierce's darting-gun. Length, 14 inches. Norwich, Connecticut, 1883. 56377. Patented, manufactured, and presented by Mr. Junius A. Brand. Substantially the same as the other kinds of Brand's lances, with the exception that, as it is driven into the whale when the gun is darted by hand, wings are not used.

ROCKETS.

ROCKET AND BOMB-SHELL.

Used in connection with the rocket-gun. This projectile consists of a cast-iron shell with three cutting edges, a brass rocket-shell, and an iron loop-extension screwed to rear of the rocket. The bomb and rocket are intended to be connected with a breech-piece. The shell has been detached to show the toggle, which is fastened by two links to the projecting end, or shoulder, of the rocket, and, when used, is entirely inclosed in the body of the shell. When the bomb explodes the toggle and chain are released, and become fastened in the blubber or flesh, preventing the apparatus from being withdrawn. An iron link or loop, with two arms, is adjusted to the loop-extension, or double shank. The end of the iron-strap is made fast to this link. Length of shank and rocket, 66 inches; length of toggle, 9¾ inches; length of bomb, 15¼ inches; shank, rocket, and toggle, 56373; bomb, 56374; iron-strap, 56375. The bomb may be loaded with an explosive compound which is ignited by the rocket-shell. The fire is communicated to the combustible material in the rocket-chamber by means of a pistol attached to the gun. The issue of gas from the rear of the rocket propels the apparatus.

APPARATUS FOR MANIPULATING DEAD WHALES, CUTTING OFF THE BLUBBER, BOARDING, MINCING, AND TRYING-OUT.

BLUBBER TACKLE.

ROPES AND BLOCKS FORMING A PURCHASE FOR HOISTING IN THE BLUBBER.

CUTTING-BLOCKS.

Two upper blocks, one guy-block, and one lower block. Lower block *strapped* with rope. The earliest method adopted for "strapping" the lower block, and in use at present on the majority of the vessels, but has been done away with on others by the improved chain strap; when used the rope strap with eye, or "grommet," is passed through a hole cut for the purpose in the blubber, and made fast (toggled) with the blubber-fid, which is inserted in the eye, thus fastening the blubber to the cutting-tackle by means of which it is hoisted on board. Lower block 18 by 12 by 10 inches; upper blocks 18 by 12 by 6 inches; guy-block, 13 by 9 by 6 inches. New Bedford, Massachusetts, 1876. 56861. E. B. & F. Macy. Total weight of blocks, hook, and toggle, 231 pounds.

CUTTING-FALLS.

Manila. Rove through upper and lower blocks. Length, 23 fathoms. Washington, D. C. 51712. U. S. Fish Commission.

WHALEMAN'S HOOKS.

USED IN BOAT.

BOAT-HOOK.

Hook round bend, with projecting spur. Length, 8 feet. New Bedford, Massachusetts, 1876. 25614. Gift of Humphrey S. Kirby. Used in the whale-boat as an ordinary hook of this kind.

USED ON VESSEL.

LINE-HOOK.

Iron shank with four branching round bend hooks, and spruce pole. Length, 15 feet. New Bedford, Massachusetts, 1876. 25924. Gift of E. B. & F. Macy. An implement carried on a whaling-vessel, and used principally from the deck for taking the end of the tow-line from the hands of the officer of the boat, in order that the whale may be hauled alongside and made fast.

LARGE BOAT-HOOK.

Hook round bend, with projecting spur, and large socket with sleeve, for pole. Length, 14 feet 6½ inches. Fairhaven, Massachusetts, 1883. 56424. U. S. Fish Commission. Used from the vessel when cutting-in a whale for hauling upon and

LARGE BOAT-HOOK—Continued.

removing the lines from the harpoons which are fastened to the whale. As much of the tow-line as can be saved in this way is subsequently used for making iron-straps, warps, &c.

LARGE-RING BOAT-HOOK.

Hook, iron, round bend, barbless. Projecting iron spur. Socket with extended sleeve. Pole, spruce; small iron ring for bending on a rope. Total length, 15 feet 7 inches. New Bedford, Massachusetts, 1876. 25926. Gift of E. B. & F. Macy. An implement used on a whaling-vessel when cutting-in a whale for pressing upon the *back* of the blubber-hook to direct the point into the hole made in the blubber of the first blanket-piece, and for hauling pieces of blubber about deck.

LARGE BLUBBER-HOOK.

Made in blacksmith shop of best refined iron. Loose ring for shackling to block. When used a rope is bent into the small ring at the *heel* of the hook, by which one of the officers directs the point into the hole cut into the blubber. A small hook, used probably on a schooner. A large and stiff ship would need a much larger and stronger hook, as the hooks are sometimes broken. Length, 26 inches. New Bedford, Massachusetts, 1876. 56861. E. B. & F. Macy. The blubber-hook is used principally for raising the blanket-piece, which is the initial point for stripping off the blubber. The pectoral fin and *head* of blanket-piece having been hoisted up "two blocks," and the first piece boarded, the hook is detached from the block, and the strap and toggle, if a *rope strap*, or the chain alone, if a *chain strap*, used in hoisting in the balance of the blubber. (*Vide* Blubber Tackle.)

SMALL BLUBBER-HOOK.

A kind of a "roustabout" hook not in general use, but may be employed in handling blanket-pieces in the hold of the vessel, in clearing the hatch when blocked with blubber, as well as in hauling the junk aft when it is to be *lashed;* hence the name "junk-hook," which is sometimes applied. Iron; small iron ring for bending on a rope when hauling the blubber. Length, 9 inches. 57725. New Bedford, Massachusetts, 1882. U. S. Fish Commission. Also known as the "lip-hook," and used by some right-whalemen for *hooking up* the lip of the whale when about to reeve the main line for "towing-in."

FIN-CHAIN HOOK.

A kind of hook, familiarly known as the "lobster claw," from its resemblance to the claw of the lobster; "fin-chain hook" from

FIN-CHAIN HOOK—Continued.

the manner in which it is used, and "ring-hook," from the peculiar shape of the bend. Large ring for shackling to lower cutting-block; small ring at *back* of hook for a laniard by means of which the hook is guided or "pointed" in the direction required. Length, 15 inches. Weight, 32 pounds. New Bedford, Massachusetts, 1882. 57726. U. S. Fish Commission. Some whalemen prefer it for *hooking* into the fin-chain, and it may, in fact, be used as an ordinary blubber-hook. It is capable of withstanding great *strains*, and its peculiar ring-shaped bend affords a tenacious *grip*.

FLUKER.

Slender spruce pole and a conjoined condemned lance shank and socket, the shank being bent, forming a round bend, until its point is directly opposite the socket. Rare. Length, 11 feet. New Bedford, Massachusetts, 1882. 55817. U. S. Fish Commission. An implement used on a whaling-vessel for passing a rope attached to one end of the fluke-chain around the *small* for fastening the whale to the ship prior to cutting off the blubber, the process being known as "fluking a whale." Obtained from the whaling brig "Varnum H. Hill," of New Bedford, Massachusetts. Manufactured at sea.

HOOKS USED IN CUTTING BLACKFISH.

BLACKFISH BLUBBER-HOOK.

A long, slender hook, with a broad tread in bend and a stiff eye. Used at sea for removing the blubber from blackfish. Provincetown, Massachusetts, 1882. 57705. U. S. Fish Commission.

WHALEMAN'S SPADES.

SCARFING AND LEANING.

NARROW CUTTING-SPADE.

Blade, cast-steel; short shank and socket, wrought-iron. Pole, spruce. Length, 15 feet. New Bedford, Massachusetts, 1876. 25928. Gift of E. B. & F. Macy. The smallest spade used when cutting-in a whale, for what is technically termed *scarfing*. Also known as the *thin boat-spade*.

NARROW CUTTING-SPADE.

Head, steel. Shank and socket, wrought iron. Nickel-plated. New. Total length, 12 feet, four inches. Fairhaven, Massachusetts, 1882. 55808. Manufactured and presented by Luther Cole. Used for *scarfing* (cutting the blubber into helical strips).

WIDE CUTTING-SPADE.

Head, cast steel; socket, wrought iron. Pole, wood. Blade has curved edge. Rare. Total length, 15 feet 6 inches. New Bedford, Massachusetts, 1876. 25008. Gift of J. H. Thomson. A kind of spade used for "leaning-up," that is, severing the pieces of flesh which adhere to the blubber when cutting-in a whale.

MORTICING HOLES IN BLUBBER.

HALF-ROUND SPADE.

Blade, in the shape of a gouge, cast-steel; shank, wrought iron. Pole, spruce. Length of spade, 15½ inches. Length of spade and handle, 15 feet 4 inches. New Bedford, Massachusetts, 1876. 25927. Gift of E. B. and F. Macy. The half-round spade is used by sperm-whalemen for making a large hole in the blubber for the blubber-hook. It is also used, though seldom, from the waist of the vessel, for making the holes in the blanket-piece which are used in fastening the blubber to the cutting-tackle.

DECAPITATING THE WHALE.

HEAD-SPADE WITH WOODEN HANDLE.

Large, heavy head, cast-steel; strong wrought-iron shank, 1¼ inches in diameter, with socket and sleeve riveted in three places to a stout wooden handle. Total length 10 feet. New Bedford, Massachusetts, 1882. 55813. Gift of Jonathan Bourne. Used for cutting through the bone when decapitating the whale.

IRON HEAD-SPADE.

A very heavy head-spade with steel head; wrought-iron handle served with spun-yarn, with a rigid eye in extreme end for a rope. Length 10 feet 10 inches. New Bedford, Massachustts, 1882. 55867. Gift of Jonathan Bourne. Rare; usually wooden poles. Employed in cutting the head-bone when decapitating the whale.

CUTTING SLIVERS.

SLIVER SPADE.

The widest *cutting-in* spade used by whalemen. Blade, cast-steel; short shank with socket and wooden handle. Length of spade 21 inches. Total length, including handle, 13 feet 7½ inches. Provincetown, Massachusetts, 1882. 55805. U. S. Fish Commission. A kind of spade used when cutting off the head of a whale for severing the connecting pieces of flesh, which are technically termed "slivers." It may also be used as a blubber-room spade by inserting a shorter handle.

CUTTING OUT THROAT-BONE, ETC.

THROAT-SPADE.

Head, or blade, cast steel; shank and socket, wrought iron. Shank round. Pole, spruce. Total length, 15 feet. New Bedford, Massachusetts, 1876. 25925. Gift of E. B. & F. Macy. A spade used for cutting a passage for the *head-strap*, in order that the head of the right whale, or bowhead whale, may be hoisted on deck, and for getting out the throat-bone (baleen). This kind of spade may be made with a round or flat shank, which should bend easily.

CUTTING BLUBBER ON DECK AND IN THE BLUBBER-ROOM.

WHALEMAN'S DECK-SPADE.

Blade, cast steel; handle, spruce. Length, 6 feet. New Bedford, Massachusetts, 1882. 57701. Gift of Thomas Knowles & Co. May be used with its present handle for cutting up blubber on deck when the main hatch is blocked, and with a longer handle as a *pot-spade* for "spading pots," to prevent refuse pieces adhering to the sides and bottoms of the pots when trying-out oil.

BLUBBER-ROOM SPADE.

Blade, cast steel; short shank, with socket for handle. Handle, wood, with cross-piece at upper end. Total length, 3 feet 11 inches. Width of blade, 7½ inches. New Bedford, Massachusetts, 1876. 57700. U. S. Fish Commission. A wide spade used in the blubber-room for reducing the blanket-pieces to horse-pieces prior to rendering the oil.

TOWING THE WHALE.

REEVING TOW-ROPE THROUGH THE LIPS.

THICK BOAT-SPADE.

Head, cast-steel; shank and socket, wrought iron. Pole, wood. New. Nickle-plated. Length, 12 feet. Fairhaven, Massachusetts, 1882. 55810. Manufactured and presented by Luther Cole. Carried in the boat, and used for making holes in the lips of the whale for reeving the tow-rope. Formerly used for stopping a running whale by severing the tendons at the junction of the caudal fin and body.

AXES.

DECAPITATING THE WHALE.

HEAD-AXE.

Common axe used by boatsteerers in cutting the bone when decapitating a whale. Length, 32½ inches. New Bedford, Massachusetts, 1876. 25913. Gift of E. B. & F. Macy. Sometimes used instead of the head-spade, in smooth weather.

CHAINS.

HOISTING IN BLUBBER.

FIN-CHAIN.

Heavy chain with large triangular loose link, or "ring" at one end, and small "ring" at the other. Length, 15 feet. New Bedford, Massachusetts, 1882. 57721. U. S. Fish Commission. Common to all whaling-vessels. Used for raising the fin and the "head" of the first blanket-pieces. Some of these chains have a loose ring shackled to the chain for the blubber-hook.

HEAD-CHAIN.

HOISTING IN HEAD OF WHALE (CASE AND JUNK).

CASE-CHAIN.

Case-chain, technically termed the "head-strap," "case-strap," or "junk-strap," employed in the sperm-whale fishery. Length, 7 feet. New Bedford, Massachusetts, 1882. 57722. U. S. Fish Commission. The whale having been decapitated, and the head subdivided, if a large whale, into two sections—the "junk" and "case"—one end of the chain is rove through a hole made in the case or junk; the other is passed through the bight, or loop, and made fast to the lower block of the blubber-tackle. The case, which contains the spermaceti, may be hoisted in a vertical position, the lower end remaining in the water, and its contents bailed over the side of the ship, or it may be hoisted on deck. The entire head of a small whale may be also hoisted in with this style of chain; hence "head-strap." This chain is smaller than those in general use.

TOGGLES, OR FIDS.

BLUBBER-TOGGLE.

The toggle, or fid (57724), made of hard wood, was formerly in general use on all American whaling-vessels, and is used to a certain extent, on many of them at present. This may be included among the earliest implements that have been steadily employed in this fishery. Recently, however, the improved method of strapping the lower block of the blubber-tackle has rendered the fid useless on the vessels which have adopted the new style. Notwithstanding this, the majority of vessels usually carry the fid, to be used if necessary, and more especially the Provincetown schooners, which use this implement altogether. Length, 24 inches. New Bedford, Massachusetts. 57724. Gift of Jonathan Bourne.

FISHERIES OF THE UNITED STATES.

HOISTING IN THROAT.

THROAT-CHAIN TOGGLE.

Iron toggle and chain formerly used for hoisting in the throat of the right or bowhead whale. Length of chain, 5 feet; length of toggle, 2 feet 6 inches. New Bedford, Massachusetts, 1882. 57723. U. S. Fish Commission. The head of the whale having been cut off, a hole is made in the throat, the toggle is inserted—technically, "dropped"—and by means of the blubber-tackle the throat is hoisted on deck.

WHALEMAN'S KNIVES.

BOARDING BLUBBER.

KNIVES USED FOR SUBDIVIDING THE MAIN PIECE INTO SMALLER SECTIONS WHEN HOISTING IN THE BLUBBER.

BOARDING-KNIFE WITH SHEATH.

Blade, steel; double-edged. Socket, iron. Handle, wood, with Turk's head to prevent the hand of the manipulator from losing its *grip* when the instrument is oily. Sheath, two pieces of wood seized with twine. Length of knife and handle, 60 inches. Length of sheath, 31 inches. New Bedford, Massachusetts, 1876. 25676. Gift of W. H. Cook & Co. Used in *boarding* blubber.

BOARDING-KNIFE.

Blade, cast steel; two-edged, with socket for handle. Handle, wood, turned; with cross-piece. Total length, 5 feet. Middletown, Connecticut, 1876. 26008. Gift of A. R. Crittenden. Used when cutting-in the whale for "boarding" the blubber.

BOARDING-KNIFE.

Blade, cast steel; double-edged; handle, wood, with cross-piece, and Turk's head. A boarding-knife with a short blade, evidently broken off, with point sharpened, used in cutting up the ambergris taken by the bark "Falcon." Total length, 4 feet 4 inches. New Bedford, Massachusetts, 1882. 55797. Gift of Thomas Knowles & Co. (Owners of the bark "Falcon"). Stamped, "J. Howard" (Manufacturer).

BOARDING-KNIFE.

Blade, cavalry saber, with hilt. Handle, wood, with cross-piece at end. Total length, 4 feet 3 inches. New Bedford, Massachusetts, 1882. 55798. Gift of Thomas Knowles & Co. Used for cutting the main blanket-piece, as it is rolled from the whale, into small sections, to be lowered into the blubber-room.

BOARDING-KNIFE.

Blade, navy cutlass with brass hilt; handle, turned wood, with knob on the end. New. Total length, 4 feet 4 inches. New Bedford, Massachusetts, 1882. 55868. U. S. Fish Commission. Manufactured by James Barton. Used on a whaling-vessel for cutting the blubber into sections, as it is unwound from the whale, in order that they may be lowered into the blubber-room.

MINCING BLUBBER.

MINCING BY HAND.

MINCING-KNIFE.

Blade, cast steel; back-frame, iron, slotted and riveted to blade. Handles, hard wood; ferules, steel. New. Sheath, wood, saturated in oil. Longitudinal slot for blade of knife. Three holes for rope beckets. Total length, 36 inches; length of blade, 24 inches; width of blade, $3\frac{1}{2}$ inches; length of sheath, 25 inches. New Bedford, Massachusetts, 1876. 25912. Gift of E. B. & F. Macy. Used for mincing small pieces of blubber (horse-pieces) in order that the oil may be more readily extracted when boiled.

MINCING-KNIFE AND SHEATH.

Blade, cast steel; worn out in service by being repeatedly sharpened. Sheath, wood, slotted for blade, and pierced with two holes for beckets. Length of knife, $36\frac{1}{2}$ inches; length of sheath, $25\frac{1}{2}$ inches. Knife, 55869; sheath, 57696. New Bedford, Massachusetts, 1882. Gift of Mackey & Pindar.

MINCING-KNIFE.

An old mincing-knife which has seen many years of service, showing the manner in which the width of blade has been reduced by frequent applications to the grindstone. Length, 36 inches. New Bedford, Massachusetts, 1882. 56849. Gift of Thomas Knowles & Co.

MINCING-KNIFE.

An old blubber-knife worn out in service and discarded, the blade having been ground down until worthless. Handles, wood; back, iron; blade, cast steel. Length, $37\frac{1}{2}$ inches. New Bedford, Massachusetts, 1882. 56850. Gift of Mackey & Pindar.

MINCING BY MACHINERY.

MINCING-MACHINE KNIFE.

A knife of peculiar shape used on some vessels instead of the hand-mincing knife, in connection with the mincing-machine, for

MINCING-MACHINE KNIFE—Continued.

slicing blubber before extracting the oil. Cast steel, holes in either end for fastening the blade to the frame. Length, 21½ inches. New Bedford, Massachusetts, 1882. 55800. Gift of Thomas Knowles & Co.

LEANING BLUBBER.

KNIVES USED IN THE BLUBBER-ROOM FOR REMOVING SMALL PIECES OF FLESH THAT HAVE ADHERED TO THE BLUBBER WHEN CUTTING IN.

LEANING-KNIFE.

Blade, steel; handle, two pieces, hardwood, riveted to shank of blade. Length, 13 inches. New Bedford, Massachusetts, 1882. 56936. Gift of Frederick S. Allen. Used by the "hold gang" of a whaling-vessel, for leaning blubber.

KNIVES USED BY SEALERS.

KNIFE, STEEL, AND SHEATH.

Case containing knife and steel. Sheath, made at sea, wood. Two pieces bound with brass hoops; leathern guard, or strap, for attaching case to waist-belt, stamped with ornamental design and initials (E. T.) of owner. Ordinary steel, handle "run in" with lead. Knife, bone handle, checkered, blade worn by sharpening. Length of case, 10 inches; length of knife, 12 inches; length of steel, 14 inches. New Bedford, Massachusetts, 1882. 56881. Gift of L. & W. R. Wing. Used by the "skinners" (men whose duty it is to skin or flay seals) in the seal and sea-elephant fishery. Herd's Island, Patagonia, South Georges, South Shetland, Desolation Island, &c.

RECEPTACLES EMPLOYED WHEN CUTTING-IN THE WHALE.

SCOOPING SPERMACETI FROM THE WATER.

SCOOP-NET.

Net made of strips of wood and spun-yarn, seized to a pole with spun-yarn. Handle and bow of net broken during transportation. Length, 14 feet. New Bedford, Massachusetts, 1882. 57697. U. S. Fish Commission. A kind of net carried on sperm-whaling vessels, and used during the process of cutting-in the whale, when severing the head, for scooping up small portions of spermaceti which float aft from the roots of the case—*skimming slicks*. Known to the Provincetown whalemen as "Granny-scratches." Obtained from whaling-brig "Varnum H. Hill," of New Bedford, Massachusetts. Has been used, as indicated by the small pieces of spermaceti adhering to the netting.

BAILING THE CASE.

CASE BUCKET.

Oak staves, bound with three iron hoops; bottom, one piece of wood, conical. Rope bail with leather ears. Length, 24½ inches; depth, 16 inches; diameter of top (inside), 9⅞ inches; diameter of bottom (inside), 6½ inches. New Bedford, Massachusetts, 1882. 55801. U. S. Fish Commission. Before the introduction of the improved windlass on whale-ships, it was impossible to hoist the head of a large sperm-whale on deck; the head was dissected while in the water and hoisted by sections, and the *case* was hauled up to the gangway vertically with its base uppermost. The case-bucket was attached to a whip-tackle, and, by means of a pole, pushed or "set" into the immense reservoir of oil and fat which comprise the spermaceti, and the contents were emptied into casks and tubs on deck.

BELTS EMPLOYED TO SUPPORT THE MEN.

MONKEY-BELT.

A wide canvas belt and a rope-tail. Canvas doubled, with a rope crinkle in each end, through which one end of the rope is passed and spliced to the standing part, leaving a loop large enough to allow the belt to be properly adjusted about the waist of a man. New. Length, 28 feet. New Bedford, Massachusetts, 1883. 57716. U. S. Fish Commission. A kind of belt worn by the man who goes overboard on the whale when cutting-in. The rope-tail is manipulated from the vessel by another man who steadies the one on the whale while engaged in adjusting the blubber-hooks, cutting off the head, &c.

STAGE-LINES.

Canvas belt and two rope-tails. Canvas doubled and stitched along the center. Two grommets, one in each end, through which the rope-tails (1¾ inch rope) are rove and spliced. Ends of ropes whipped. Length, 18 feet. New Bedford Massachusetts, 1882. 57713–4. Gift of Jonathan Bourne. Made fast to the main rail of the vessel and used to prevent the officers falling overboard when cutting in the whale.

REEVING CHAINS THROUGH BLUBBER.

WHALEMAN'S NEEDLE.

Hard wood; conical; recessed at large end. Rope becket for bending on small line. Length, 26 inches. Provincetown, Massachusetts, 1882. 57707. Gift of Stephen Cook. Used by sperm whalemen for reeving a small rope through a hole made in the *case* or *junk*, by means of which the head-chain or junk-chain may be hauled through.

MANIPULATING BLUBBER-HOOKS.

BLIND BOAT-STEERER.

Handle, wood. Jaws, iron, resembling those of a boom, or gaff, of a sailing-vessel. Length, 7 feet 5 inches. Provincetown, Massachusetts, 1882. 55803. Gift of Mr. Stephen Cook. Used on some whaling-vessels for pressing upon the *back* of the hook in guiding the point into the hole of the blubber of the first blanket-piece. By using this instrument the boat-steerer, whose duty it is to go overboard on the whale to insert the hook, may, in rugged weather, accomplish this from the cutting-stage. Blind boat-steerer, Provincetown, Massachusetts; dog's-legs. New Bedford and Edgartown.

PLATFORMS FOR OFFICERS WHEN CUTTING-IN THE WHALE.

CUTTING-STAGE.

A kind of platform, technically called the "forward cutting-stage," upon which the second mate stands when cutting-in a whale. The position for the stage, when in use, is on the outside of the vessel, resting against and braced from the side, by the cross-pieces, and made fast to the rail by the ropes. Not much used at present, having been supplanted by an improved form known as the "outrigger stage." Length, 42 inches; width, 16 inches. New Bedford, Massachusetts, 1882. 57727. Gift of Jonathan Bourne.

TRY-WORKS GEAR.

BAILING OIL FROM THE POTS.

LONG-HANDLED BAILER.

Bailer, tin; shank, wrought iron. Length, including handle, 13 feet. New Bedford, Massachusetts, 1882. 57749. U. S. Fish Commission. Used for transferring hot oil from the try-pots to the cooler. Present style.

HANDLE OF BAILER.

Handle and yoke of bailer common to all whaling vessels, showing rudely-carved figures of whales cut into the wood by the officer of the watch, indicating the number and species of whales boiled out. Bailer, wanting. New Bedford, Massachusetts, 1828. 55809. Gift of Jonathan Bourne. Marks indicate that it has been in service on several voyages, and in the several branches of the fishery: "B. H.," Bowhead whale; "S.," Sperm Whale; "H. B.," Humpback, and "W.," the Whale, or Right Whale.

REMOVING SCRAPS FROM TRY-POTS.

SCRAP-DIPPER.

Bowl, sheet iron, perforated; shank and socket, wrought iron; pole, wood. Manufactured by sheet-iron workers. Total length, 13 feet. New Bedford, Massachusetts, 1882. 55806. U. S. Fish Commission. A kind of skimmer or colander for removing the refuse pieces of blubber, commonly known as *scrap*, from the try-pots. Formerly made of copper or brass, but at present usually of heavy tin or galvanized sheet iron. An improved form.

BLUBBER-PIKES.

HANDLING BLUBBER WHEN MINCING.

BLUBBER-PIKE.

Small iron pike with socket and pole, used on whaling vessels for handling horse-pieces during the process of mincing the blubber. Length, 5 feet 10 inches. New Bedford, Massachusetts, 1876. 25615. Humphrey S. Kirby.

BLUBBER-PIKE.

Common iron pike with spur for attaching the instrument to the pole; strengthened by a metal band. Length, 4 feet 9 inches. New Bedford, Massachusetts, 1876. 25617. Humphrey S. Kirby. From the fore-hold of a returned whaler. Used in handling blubber when trying-out.

BLUBBER-PIKE.

A single-pointed instrument, iron, attached to a rough wooden pole by means of a spur, and held by a metal band or ferule. Total length, 4 feet 9 inches. New Bedford, Massachusetts, 1876. 25617. Humphrey S. Kirby. Used on the deck of a whaling-vessel when "mincing" for transferring horse-pieces from the blubber-tub to the mincing-tub.

POT-PIKES.

STIRRING FIRES AND HANDLING SCRAP.

POT-PIKE.

A small pike, consisting of a spur, shank, socket, and pole, with a collar welded near the bend to prevent the scrap from sliding up the shank. Length of pike, 33 inches. New Bedford, Massachusetts, 1882. 57704. Gift of James Barton. Used for removing scrap from the try-pots, pitching scrap as fuel into the arches, and for stirring up the fires.

BLUBBER-FORKS.

MINCING AND TRYING-OUT.

HORSE-PIECE FORK.

A small pitch fork, all iron, with two tines. Handle parceled. Rare. Usually has wooden poles. Length, 4 feet 9½ inches. Provincetown, Massachusetts, 1882. 57703. U. S. Fish Commission. Used in the blubber-room and on deck for handling horse-pieces when mincing and trying-out.

HORSE-PIECE FORK.

Small iron fork with two tines and wooden pole. Length, 4 feet 6 inches. New Bedford, Massachusetts, 1882. 57702. Jonathan Bourne. Used for transferring horse-pieces from the blubber-tub to the mincing-tub, and for pitching horse-pieces from the blubber-room on deck.

BLUBBER-FORK.

Small iron fork with two tines, short shank and wooden handle, formerly used for pitching horse-pieces into the try-pots, but superseded by a similar instrument (25950) with longer prongs. Length, 8 feet 9 inches. New Bedford, Massachusetts, 1882. 55818. Gift of Mackey & Pindar.

BLUBBER-FORK.

Small iron fork with two tines, socket, and pole. Used at present for pitching blubber into the try-pots. Length, 5 feet 7 inches. New Bedford, Massachusetts, 1876. 25950. Gift of E. B. & F. Macy.

BLUBBER-GAFFS.

LOWERING BLUBBER IN MAIN HATCH AND HAULING BLUBBER ABOUT DECK.

BLUBBER-GAFF.

Common iron gaff with a spur at rear end for attaching the instrument to a common pole, and held fast by a metal band. Length, with pole, 4 feet 10 inches. New Bedford, Massachusetts, 1876. 57699. U. S. Fish Commission. Used when cutting-in a whale for "pointing" the blanket-pieces over the main hatch when lowering them into the blubber-room.

BLUBBER-GAFF.

Common iron gaff attached to a rough pole. Obtained from a whaling and sealing vessel, and used when lowering the blanket-pieces down the main hatch. Length, 42 inches. New London, Connecticut, 1882. 57706. Gift of C. A. Williams & Co.

LIGHTS.

TRY-WORKS LANTERNS.

BUG-LIGHT.

An open-work receptacle made of hoop-iron, formerly suspended between the try-works pipes, filled with scrap, and used as a lantern, when boiling out at night. Superseded by a glass lantern. New Bedford, Massachusetts, 1882. 57717. Gift of Jonathan Bourne.

CUTTING-IN WHALE AT NIGHT.

BUG-LIGHT.

Open-work receptacle for "scrap" made of pieces of hoop-iron; handle, broken oar served with rope-yarn. Length, 11 feet 8 inches. New Bedford, Massachusetts, 1882. 55802. Gift of Jonathan Bourne. Made on a whaling vessel, undoubtedly by the blacksmith. An apparatus which may be projected over the side of a vessel, and lashed to the main-rail, while cutting-in a whale at night, and used as a lantern; the bowl-shaped receptacle at the end being filled with scrap and ignited. It may also be used when boiling out the oil for removing the scrap.

ROPES USED BY WHALEMEN.

ON THE VESSEL.

FLUKE-ROPE.

The largest rope employed by whalemen. Formerly used for fastening the whale to the vessel, but has been almost wholly superseded by a large chain (fluke-chain). The whale having been killed, it is towed to the vessel; the fluke-rope is passed around that portion of the animal known as the *small*, the junction of the caudal fin (flukes) and body, and made fast to the vessel. Manila hemp; circumference, 8 inches; three strands. New Bedford, Massachusetts, 1882. 56390. New Bedford Cordage Company.

CUTTING-FALLS.

A kind of rope technically known as the *cutting-falls*. Full length should be 38 or 40 fathoms. Used in connecting the lower and upper blocks, forming a *purchase*, commonly known as the "cutting-tackle" or the "blubber-tackle," by means of which the blubber is unwound in spiral strips from the whale, hoisted in, and lowered into the blubber-room. Manila hemp; four strands; circumference, 5¼ inches. New Bedford, Massachusetts, 1882. 56391. New Bedford Cordage Company.

FISHERIES OF THE UNITED STATES.

GUY-ROPE.

A kind of rope used on a whale-ship. Rove through the guy-block to hold the upper blocks of the cutting-tackle stationary when suspended over the main hatch, while hoisting in the blubber. Manila hemp; circumference, 4¾ inches; number of strands, 4. New Bedford, Massachusetts, 1883. 56392. New Bedford Cordage Company.

YARN FOR TYING BUNDLES OF WHALEBONE.

BONE-YARN.

Sample of bone-yarn carried by right whalemen, and used in tying up bundles of whalebone (baleen). Made of Russian hemp, tarred; two yarns carded together; circumference, ⅞ inch. New Bedford, Massachusetts, 1882. 56395. New Bedford Cordage Company.

WHALE-LINE, WARPS, AND STRAPS.

USED IN THE BOAT.

WHALE-LINE.

A kind of rope used in all American whale-boats during the capture for fastening the whale to the boat. Manila hemp; three strands; circumference, 2 inches. New Bedford, Massachusetts, 1882. 56393. New Bedford Cordage Company.

IRON-STRAP, SHOWING EYE-SPLICE.

A piece of tow-line, technically termed an "iron-strap," which, when used, is made fast at one end to the shank of the harpoon near the socket; has an eye-splice in the other for bending on the whale-line. Manila hemp; circumference, 2 inches; three strands. New Bedford, Massachusetts, 1882. 56396. Manufactured and presented by the New Bedford Cordage Company. (Prepared by Captain Isaiah West.)

LANCE-WARP.

A sample of the smallest line employed during the capture of a whale. One end of about eight fathoms of this line is made fast to the hand-lance and its pole, the other being fastened to the boat, and used in manipulating the lance when the officer of the boat is killing the whale. Manila hemp; circumference, ¾ inch; number of strands, 3. New Bedford, Massachusetts, 1882. 56394. New Bedford Cordage Company.

LANCE-STRAP.

A piece of lance-warp showing eye-splice, intended to be fastened to the shank of the hand-lance at the socket by a round turn and splice; seized to the pole in two or three places with rope-

LANCE-STRAP—Continued.

    yarn, with an eye-splice in the other end for making fast the lance-warp. Length should be 6 or 7 feet. New Bedford, Massachusetts, 1882. 56399. Manufactured and presented by the New Bedford Cordage Company.

SHORT WARP.

    A piece of whale-line, with a bowline at one end, and crowned and the ends expended at the other to prevent the rope from unlaying (unraveling). One end of this warp is intended to be made fast to the strap of the "second iron"; the other end is bent around the whale-line with a bowline, in order that the line may be run freely when taken out by the whale. Manila hemp, 2 inches in circumference, three strands. Length, 4 fathoms. New Bedford, Massachusetts, 1882. 56397. Manufactured and presented by the New Bedford Cordage Company.

WAIFS AND FLAGS.

LOCATING THE WHEREABOUTS OF DEAD WHALES, AND SIGNALS FOR THE VESSEL.

BOAT-WAIF.

    A small flag with a grayish blue (dungare) ground and white square and compass (cotton cloth), made fast to a slender pine pole; 55 by 34 inches. 56854. U. S. Fish Commission. A kind of flag used as a signal in a whale-boat, and for *waiving* (marking) a dead whale.

BOAT-WAIF.

    Small flag with white field (cotton cloth) and dungare crescent, attached to a slender pine pole; 55 by 34 inches. 56855. U. S. Fish Commission. A kind of flag used as a signal in a whale-boat, and for *waiving* (marking) a dead whale.

NATIONAL FLAG.

AMERICAN ENSIGN.

    Ensign carried twelve years in the Hudson Bay fishery by the whaling-schooner Abbie Bradford (114.75 tons), of New Bedford, Massachusetts. 8 by 12 feet. New Bedford, Massachusetts, 1882. 57720. Gift of Jonathan Bourne. The American ensign is always carried at the mizzen peak by whaling-vessels.

IMPLEMENTS USED ASHORE.

For Scraping and Cleaning Slabs of Baleen, and by Coopers, both Ashore and at Sea, for Smoothing the Interior Surfaces of Wooden Utensils.

BONE-SCRAPERS.

BONE-SCRAPER.

Handle, wood; blade, common hoop-iron, riveted to handle. Roughly made. Length, 8½ inches. New London, Connecticut, 1882. 57072. Gift of Lawrence & Co.

SCRAPER.

Roughly-made handle, wood; half ovate blade with spur for insertion in handle. Metal ferule. Length, 8¼ inches. New Bedford, Massachusetts, 1882. 57074. Gift of Jonathan Bourne. May be used as a bone-scraper or inshave.

BONE-SCRAPER.

Handle, rough wood; iron shank; ovate frame forming the blade, common hoop-iron. Blade riveted to shank. A very old specimen. Length, 11½ inches. New Bedford, Massachusetts, 1882. 57076. Gift of Thomas Knowles & Co.

BONE-SCRAPERS.

Blades, steel, slotted vertically in handle. Handles, wood; holes in handles for small laniards. Length, 3¼ inches. New Bedford, Massachusetts, 1882. 57068. Gift of F. S. Allen. Used in cleaning whalebone (baleen).

COOPER'S INSHAVES.

INSHAVE.

Handle, turned wood. Iron frame, a true oblong ovate, with blade on forward edge, and spur for insertion in handle. Metal ferule. Length, 11½ inches. New Bedford, Massachusetts, 1882. 57067. Gift of Jonathan Bourne.

COOPER'S SMALL INSHAVE.

Handle, wood. Iron frame, oblong ovate; cutting-edge on forward part. Metal ferule. Length, 10 inches. New Bedford, Massachusetts, 1882. 57069. Gift of John McCullough.

COOPER'S LARGE INSHAVE.

Handle, turned wood. Iron frame with cutting edge and rear extension for attaching to handle. Length, 12¼ inches. New London, Connecticut, 1882. 57070. Gift of C. A. Williams & Co.

COOPER'S LARGE INSHAVE.
  Socket and shank, iron. Ovate frame with sharp cutting-edge. Length, 19¼ inches. New London, Connecticut, 1882. 57071. Gift of C. A. Williams & Co.

COOPER'S INSHAVE.
  Handle, wood; blade, steel, with spur for fastening to handle. No ferule. Old. Length, 6¼ inches. New London, Connecticut, 1882. 57073. Gift of C. A. Williams & Co.

COOPER'S INSHAVE.
  Handle, wood; frame, acute ovate, with forward cutting-edge riveted to handle. Length, 7 inches. New Bedford, Massachusetts, 1882. 57075. Gift of Thomas Knowles & Co. An old inshave used for many years on a whaling-vessel.

LOGS OF WHALING-VESSELS.

LOGS.
  Journals containing daily entries of the vessels' routine; remarks upon the weather, sky, wind, localities, and whaling-grounds visited, including latitudes and longitudes, the number of whales captured, amount of oil boiled out and stowed down, and other matters of importance which tend, in the aggregate, to make a true register of the voyages. The mates usually keep the logs, which are, in many cases, illustrated with cuts of whales and profiles of the islands passed or visited during the voyage. A figure of a sperm-whale, for example, stamped upon the page of a journal with the initials "L. B." and the figures "40," indicates that upon the day of that entry a sperm-whale, yielding 40 barrels of oil, was captured by the larboard boat. The *flukes* of a whale in a vertical position indicate that whales were seen but not captured. Half of a whale indicates that the vessel "mated," that is, entered into an agreement with another vessel to jointly capture the whale, and that she secured one-half of the prize. The "twenty-four hours" commences at 12 o'clock at night and ends at 12 midnight. Formerly the English sea-journals' day, or twenty-four hours, "used to terminate at noon, because the ship's position is then generally determined by observation; but the shore account of time is now adopted afloat."*

WHALEMAN'S LOG.
  Journal of bark Peri, Capt. E. Russell. Sailed Friday, June 29, 1854, from New Bedford, Massachusetts, for the Indian Ocean on a sperm-whaling voyage, and returned May 26, 1857. Edgartown, Massachusetts, 1882. 56865. Gift of John W. Norton.

---
* Admiral W. H. Smyth.

WHALEMAN'S LOG.

Journal of a part of a voyage made by bark Virginia, R. G. Luce, commander. Sailed from New Bedford, Massachusetts, August 15, 1855, for the Pacific Ocean. Edgartown, Massachusetts, 1882. 56866. Gift of Thomas M. Peakes.

WHALEMAN'S LOG.

Journal of ship Mary, of Edgartown, Massachusetts, Capt. G. A. Baylies. Sailed on a whaling voyage June 8, 1852. Edgartown, Massachusetts, 1882. 56867. Gift of Mrs. E. A. Gannett.

WHALEMAN'S LOG.

Journal of bark "Adeline Gibbs," Captain G. H. Baylies, of Fairhaven, Massachusetts. Sperm-whaling voyage in the Pacific Ocean, 1841–1845. New Bedford, Massachusetts, 1882. 56868. Gift of John McCullough.

WHALEMAN'S STAMP.

Redwood stamp with figure of sperm-whale (*Physeter macrocephalus*). Length, $3\frac{1}{4}$ inches. Edgartown, Massachusetts, 1882. 56869. Gift of Capt. J. E. Osborn. Made at sea by Captain Osborn, and used in stamping the figures of whales in log-books.

WHALEMAN'S LOG SLATE.

Double slate; wooden backs, hinged. Used by the mate for making rough notes which are subsequently entered in the journal. Dimensions, 17 by $17\frac{1}{2}$ inches. New Bedford, Massachusetts, 1882. 56870. Gift of Daniel Kelleher.

WHALEMAN'S SLATE-PENCIL BOX.

Common wooden box used by the mate of a whaling vessel, as a receptacle for slate-pencils. Length, $8\frac{1}{4}$ inches. New Bedford, Massachusetts, 1882. 56871. Gift of Daniel Kelleher.

WHALING-VOYAGE JOURNAL.

Journal of the ship Dauphin in verse (doggerel). Composed by Charles Murphey, third mate, on the voyage. Ship Dauphin sailed September 4, 1820, from Nantucket. New Bedford, Massachusetts, 1881. 56893. Gift of Captain Amos C. Baker.

> 'Twas one-and-twenty men we had
> This voyage to pursue,
> And a sperm-whaling we were bound
> On Chili and Peru.

MEMORANDA OF OUTFIT OF A WHALING VESSEL.

Small pamphlet with printed lists of supplies required for a whaling vessel. New. New Bedford, Massachusetts, 1883. 57020. James V. Cox.

SHIP'S PAPERS.

Copies of papers carried by whaling bark Bartholomew Gosnold, of New Bedford, Massachusetts, outward bound. Register, 57016; whalemen's shipping papers, 57017; certificate to shipping articles, 57018; crew lists, 57019; master's certificate, 57020; custom-house fees, 57021; bill of health, 57022. New Bedford, Massachusetts, 1883. Captain James V. Cox and Mr. James Taylor.

ACCESSORIES.

BELLY-BAND.

Consists of a belt and two rope-tails. Belt, braided rope, with an eye in each end, into which the ends of the two ropes are respectively spliced. Total length of belt and ropes, 11 feet. New Bedford, Massachusetts, 1882. 57715. Gift of Jonathan Bourne. Used by the men when drawing water over the side of the vessel when the ship is under way. The belt having been adjusted about the waist of a man, who stands on the chains, the free ends of the ropes are made fast to the main-chains of the ship. The man, thus supported, having filled the bucket with water, swings it up to another man, who leans over the rail. The contents having been emptied into a large deck tub, the bucket is again hove overboard and the operation repeated until a sufficient quantity of water has been obtained.

WHALEMAN'S BOOT-JACK.

Yellow pine; rest, Spanish cedar. Made on a whaling vessel. (Scrimshaw). Length, 13 inches. New London, Connecticut. 1882. 56883. Gift of Lawrence & Co.

WHALEMAN'S "BELL."

Wood, oak; three pieces; evidently made of an oil-cask stave. Center-piece with a projecting handle; short side-pieces, seized to the handle-piece with a leathern thong. Total length, 11 inches. New Bedford, Massachusetts, 1882. 56882. Gift of Captain Henry Clay. Obtained from the whaling schooner Golden Eagle, known as the "cracker," "rattler," and "Nantucket bell." One of the oldest implements employed on whaling vessels, and used at present on some of the Provincetown and New Bedford schooners, whose crews retain the customs and habits of the early whalemen. The full-rigged barks and ships, however, have discarded the "clapper," and in its place use the bell common to all first-class vessels. When the time arrives for relieving the man at the wheel, he calls another member of the watch by *rattling the bell.*

WHALEMAN'S HAND-CUFFS.
Pair of hand-cuffs (with key), connected by two loose links and a swivel. Length, 9.2 inches. New Bedford, Massachusetts, 1882. 56887. U. S. Fish Commission. Common to all whaling vessels for enforcing discipline, manacling insubordinate, pugilistic, or drunken members of the crew, and deserters if caught.

PAIR OF SKATES.
Foot-rest, hard wood; runners made from old files, fastened to the rear ends of skates with common wood-screws, the projecting ends of the latter secured to the boot-heel when used. Heel-straps, common leather; wanting front straps. Length, 12 inches. New London, Connecticut, 1882. 56885. Gift of Lawrence & Co. Made at sea, and used by sea-elephant hunters.

REEL AND LOG-CHIP.
Reel, wood; iron axle, projecting wooden handles. End-pieces bucked with iron. Log-chip, common form, triangular; base armed with lead. Length of reel, 25 inches, 57081; size of chip, 6 by 6 inches, 57080.

ANIMAL-TRAP.
Common steel trap. "Hawley & Norton, No. 1" (New York), used by whalemen in the Arctic regions for the capture of foxes and other small animals for their fur. Length, 31 inches, including chain. New London, Connecticut, 1882. 57799. Gift of C. A. Williams & Co.

MAIN-ROYAL POLE.
Pine, made expressly for its present use, showing its connection with the lookout bows. Those in active service are made of spruce or southern pine. Length, 8 feet 6 inches. Washington, D. C., 1883. 57719. U. S. Fish Commission.

MINUTE-GLASS.
Common glass formerly used in connection with the log. Old. Four uprights, seized at top and bottom with twine. Height, 4¾ inches. New Bedford, Massachusetts, 1882. 57082. Gift of Jonathan Bourne.

LOOKOUT BOWS.
Two iron rings; parceled; shackled to main-royal pole. Inside diameter, 18 inches. New Bedford, Massachusetts, 1882. 57718. U. S. Fish Commission. A support for the men when on the lookout for whales. The men climb up, and by means of the rigging lower themselves into the bows, and standing upon the cross-trees, support themselves by grasping the rings or rigging.

MARINE-GLASS BAG.
  Garnet plush, with canvas shoulder-strap, made at sea, and used as a receptacle for the marine glass. Obtained from a sealing vessel. 6 by 8 inches. New London, Connecticut, 1882. 57710. Gift of Lawrence & Co.

MARINE-GLASS CASE AND BAG.
  Bag, canvas, with rope strap; ordinary marine-glass case, small size. Used by officers and boat-steerers when at the mast-head on the lookout for whales. 7 by 7¼ inches. New Bedford, Massachusetts, 1882. 57709. U. S. Fish Commission.

MARINE-GLASS BAG.
  Small canvas bag, with rope shoulder-strap, used by the men on the lookout for whales as a receptacle for the marine glass. 7¼ by 9 inches. New Bedford, Massachusetts, 1882. 57708. Gift of Thomas Knowles & Co.

WHALEMAN'S BUNG-THIEF.
  Wood, one piece. Gouged out. Leather bail and codline lanyard. One side loaded with lead, for submerging the cup; opposite side chamfered for convenience in drinking. Made at sea. Length, 12½ inches. New Bedford, Massachusetts, 1882. 56873. Gift of Loum Snow & Son. A drinking cup carried on the deck of a whaling vessel for obtaining fresh water, its proper place being on or about the fresh-water tank. If the cask rests upon its bilge, the "thief" may be inserted through the bung-hole, and the supply of water obtained, or if the cask stands upon one end, as is often the case, the "thief" is dropped through a square-cut hole in the head.

ESSENCE OF SPRUCE.
  A pint bottle, colorless glass, containing essence of spruce. Included in the outfit of whaling vessels for making spruce beer which contains water, molasses, and essence of spruce mixed; placed in a cask and fermented. Height, 10½ inches. New Bedford, Massachusetts, 1882. 56876. Gift of Daniel Kelleher.

WHALEMAN'S LEG-IRONS.
  Small iron rod with two loose shackles for the ankles, and one loose shackle for chaining the victim to some stationary object. Length, 11 inches. New Bedford, Massachusetts, 1882. 56878. U. S. Fish Commission. Used on insubordinate members of the crew, who, when manacled, are placed in the run, or between decks in the blubber-room, and kept on bread and water until they are willing to comply with the rules of the ship. Not often used, but always carried on whaling vessels.

WHALEMAN'S SHOES.
  A kind of brogan worn by whalemen; No. 11, pegged. New Bedford, Massachusetts, 1882. 56879. U. S. Fish Commission.

WHALEMAN'S SHIP-BREAD.
A kind of ship biscuit, usually known as "hard-tack," common to the majority of sea-going vessels. Weight, 3¼ ounces. New Bedford, Massachusetts, 1882. 56880. Gift of J. T. Buttrick (Manufacturer). Baked expressly for a whaling voyage, made of common wheat flour and cold water, without salt or other ingredients; kneaded, cut into shape, and perforated by machinery driven by steam, and baked on soap-stone in a rotary oven; packed in air-tight ten-barrel casks when stowed away in the ship's hold.

SCRIMSHAW WORK AND CURIOSITIES.

WHALEMAN'S TROUSERS.
Pair of canvas pants, old style, with fly, showing rent made by the jaws of a shark. Worn by N. N. Cook, schooner "Belle Isle," Provincetown, Massachusetts, when bitten by a shark, on February 22, 1841, in Samana Bay, West Indies, while discharging the duties of boat-steerer on the whale, adjusting the blubber-hook when cutting-in. Full length, 37 inches; waist, 28 inches. Provincetown, Massachusetts, 1882. 56898. Gift of N. N. Cook.

SCRIMSHAWED SHIP.
Model of whale-ship carved in relief; hull, sails, masts, and spars, wood. All sails set. Hull painted black; sails, white. Waves represented by putty, painted. Box frame. Size of frame, 26¼ by 13½ inches. New London, Connecticut, 1882. 57059. Gift of Lawrence & Co. Made at sea by a whaleman.

CANE.
Walking-stick made of whalebone (baleen) by a whaleman at sea (*Scrimshaw* work). Heart, several pieces, wrapped spirally with strips of baleen, and *wormed* with cord of the same material. Three Turk's head, baleen, at top, bottom, and center. Length, 33 inches. Edgartown, Massachusetts, 1882. 56897. Gift of J. W. Coffin.

WHALEMAN'S BANJO.
Body, an old tin fruit or vegetable can, over one end of which a piece of porpoise skin is strained and held with spun-yarn. Neck, hickory; sound-board and pegs, pine. Three strings, common wrapping twine, waxed. Length, 20 inches. Provincetown, Massachusetts, 1882. 56872. Gift of George O. Knowles. Made on board schooner "Quickstep," of Provincetown, Massachusetts, by a negro (Portuguese) whaleman. It is said by his shipmates that the manufacturer "discoursed most excellent music" upon this rudely constructed instrument.

WHALEMAN'S PITCH-DIPPER.
A utensil made by attaching a pine wood handle to a common periwinkle shell (*Sycotypus canaliculatus*), and used for handling pitch when paying deck-seams. Length, 10 inches. New London, Connecticut, 1882. 56875. Gift of Lawrence & Co.

JIG-TACKLE.
Forward part of jig-tackle, grafted and painted. Crupper-like arrangement at forward end for shipping over the bow-chock; ivory block at other end. After end wanting. Length, 4 feet 4½ inches. Noank, Connecticut, 1880. 57065. Made and presented by Captain H. C. Chester. A tackle in which many of the boat-steerers take great pride. Used to prevent the whale-boat from chafing when on the cranes.

JIG-TACKLE.
Two parts, the forward and after ends accompanied by the falls. The ropes forming both parts are unla'd; each strand neatly covered with canvas and braided into round "sennit." The strap forming the forward part has in one end a crupper-like arrangement, covered with leather to prevent chafing, which fits over the bow chocks of the whale-boat, and a small eye in the other end for a block. The strap forming the after end has at one extremity an eye for a block, and a wooden cleat at the other, which is made fast to the bearer. The two parts when in use are hauled together and hold the whale-boat when transported on the vessel in its proper position; both parts painted blue. Chafed and worn in service. Falls, 9-thread manila. Length of forward part, 20 inches; length of after part, 60 inches. New Bedford, Massachusetts, 1882. 57728. Gift of L. & W. R. Wing.

MAN-ROPES.
Rope, grafted with cotton cloth and painted white. Man-rope knot at one end, *painted* at the other. One pair. Length, 9½ feet. New Bedford, Massachusetts, 1882. 57060. Gift of L. & W. R. Wing.

MAN-ROPE STANCHIONS.
Bone, cut from the *pan* of the sperm-whale's jaw. Known to whalemen as ivory. Feet square-cornered to ship in sockets in vessel's rail. Eye in upper end for man-rope. One pair. Length, 15 inches. New Bedford, Massachusetts, 1882. 57062. Gift of Daniel Kelleber.

CHEST-BECKETS.
Rope strands braided; painted green. Flemish eyes; cross-bar passing through eyes and knotted at each end. One pair. Length, 11 inches. New Bedford, Massachusetts, 1882. 57061. Gift of L. & W. R. Wing. Made at sea by a whaleman and used as beckets or handles for clothes-chest.

SPLICING-FID.
    Bone-fid made from *pan* of sperm-whale's jaw. Usually called ivory by whalemen. Length, 11½ inches. New Bedford, Massachusetts, 1882. 57063. Gift of Daniel Kelleher. *Scrimshaw* work, made at sea by a whaleman. Ordinary fid for splicing rope.

COG-WHEEL.
    A ratchet wheel made of "ivory" (*pan* of sperm-whale's jaw). Iron shaft. Length, 4¾ inches. New Bedford, Massachusetts, 1882. 57064. Gift of Daniel Kelleher. *Scrimshaw* work, made at sea by a whaleman for some mechanical device.

SABER.
    A common cavalry saber obtained from a whaling vessel. Length, 37 inches. New London, Connecticut, 1882. 56886. Gift of Messrs. Lawrence & Co. Used on board ship in the manufacture of boarding-knives, etc.

MACHETA-KNIFE.
    Thick, heavy blade, with a wide, curved point. Handle horn. Length, 27¾ inches. New London, Connecticut, 1882. 56874. Gift of Lawrence & Co. Knives of this character are used in the West Indies for cutting sugar-cane; in Mexico, Central America, and tropical South America as an axe for felling trees, as well as for defensive and offensive weapons. A similar form is also used by rubber-hunters. Imported by whalemen. Used on board whaling vessels in the manufacture of knives, etc.

WAR-CLUB.
    Two-edged sword, cocoanut wood (*Cocos nucifera*), armed on the sides of blade with sharks' teeth (genus allied to *Carcharias*). Teeth seized with coir. Becket in handle, coir. Length, 28 inches. Edgartown, Massachusetts, 1882. 56895. Gift of J. W. Coffin. Obtained in South Pacific and brought home as a "curio" by a whaleman.

EAR-BONE OF CALF-WHALE.
    Ear-bone of sperm-whale calf (*Physeter macrocephalus*), brought home as a *curio*. Fairhaven, Massachusetts, 1882. 56894. Gift of Girard S. Robinson.

KANAKA-LINE.
    Heart, coir. Surface, plaited vegetable fiber. Edgartown, Massachusetts, 1882. 56896. U. S. Fish Commission. Brought home by a whaleman. Said to be used by natives of the Sandwich Islands as an ornament for the person by affixing shells and small teeth, and in the manufacture of baskets, etc. Called by whalemen "Kanaka-line."

MODEL OF ESKIMO SALMON-SPEAR.
Model of a kind of salmon-spear used by the Eskimo of Hudson Bay. Wooden handle, with a central brass barbless spear, and two diverging wooden prongs, with bent tacks as barbs. Length, 13¾ inches. New Bedford, Massachusetts, 1882. U. S. Fish Commission. This model was made by a native at the request of a whaleman, and is said to be a correct representation of the original, with the exception of the tacks, wooden prongs, and brass spear—these parts being usually made from bone. Obtained from the crew of whaling brig "George and Mary."

ESKIMO SPOON.
A domestic utensil, made from the horn of the musk-ox, used by Eskimo of Hudson Bay as a spoon in eating soup. Length, 2⅜ inches. New Bedford, Massachusetts, 1882. 63145. U. S. Fish Commission. Obtained from crew of whaling brig "George and Mary."

ESKIMO HUNTING-CASE, BOWS, AND ARROWS.
Case, deer-skin; one bow has end-pieces made of the ribs of the deer, and center-piece made of walrus tusk; the other bow is made of ribs of deer and wood. Thongs made from sinew of the deer. Bows and case have been used. Arrows new. Length of case, 34¼ inches. New Bedford, Massachusetts, 1882. 68127. U. S. Fish Commission. Used by some of the tribes in hunting deer, walrus, musk-ox, seal, bears, partridges, etc. Obtained from crew of whaling brig "George and Mary."

ESKIMO PIPE.
Small bowl of the Chinese form. Stem, two pieces of wood, bound with two small brass hoops or rings. Ornamented with glass beads, red, white, and blue, and pendant shark's-teeth. Length, 7 inches. New Bedford, Massachusetts, 1882. 68140. U. S. Fish Commission. Obtained from crew of whaling brig "George and Mary."

ESKIMO SHOES.
Pair of infant's shoes worn indoors, seal-skin, sewed with thread made from the sinew of the backbone of the deer. Eskimo, Hudson Bay. Length, 4½ inches. New Bedford, Massachusetts, 1882. 68142. U. S. Fish Commission. Obtained from crew of whaling brig "George and Mary."

ESKIMO THREAD.
Sinew of the backbone of the deer, used by Eskimo of Hudson Bay in making clothing, shoes, thongs for bows, tying up the hair, etc. Length, 22 inches. New Bedford, Massachusetts, 1882. 68144. U. S. Fish Commission. Obtained from crew of whaling brig "George and Mary."

FISHERIES OF THE UNITED STATES.

PIECE OF BLACKSKIN.

A small section of tough skin, termed "white-horse," cut from the "bonnet" of a right-whale, invested with crustacean parasites, the "barnacles" of the whalemen; shows the ravages of cockroaches while on the vessel. Brought home as a curio. Length, 8½ inches. New Bedford, Massachusetts, 1882. 57094. U. S. Fish Commission.

SHELL HOOK.

Shank made from the hinge of a pearl-bearing shell (*Avicula margaritifera*); hook portion of border of probably the same species, made fast to shank with a *seizing* of some vegetable fiber. Length, 5 inches. New Bedford, Massachusetts, 1882. 68139. U. S. Fish Commission. Called "Kanaka hook" by whalemen, the word "Kanaka" being vaguely and comprehensively applied to articles obtained from the islands of the South Pacific.

CHILD'S STOCKINGS.

Seal-skin, sewed with thread made from the sinews of the back of the deer. Made and used by Eskimo, Hudson Bay. Length, 5 inches. New Bedford, Massachusetts, 1882. 68143. U. S. Fish Commission. Obtained from crew of whaling brig "George and Mary."

EYE-PROTECTORS.

Wood; two longitudinal slits; straps made from red cloth, used by Eskimo, Hudson Bay, and American whalemen, to shield the eyes from the glare of sun and snow. Length, 4⅜ inches. New Bedford, Massachusetts, 1882. 68141. U. S. Fish Commission. Obtained from crew of whaling brig "George and Mary."

SNOW-KNIFE.

Long blade, said to be made from a whaleman's boarding-knife, the original having been made from a navy cutlass. Handle, walrus ivory. Length, 17½ inches. New Bedford, Massachusetts, 1882. 68125. U. S. Fish Commission. Obtained from one of the crew of whaling brig "George and Mary." Made and used by Eskimo, Hudson Bay, for cutting out blocks of snow in building igloos, as well as for cutting walrus meat.

IDOL.

A species of the gourd family (*Leginaria vulgaris*), obtained by a whaleman from a small island near the coast of New Guinea, East Indies. As near as the captain of the vessel could understand from the pantomimic gestures of the natives, it was worshiped as an idol, and represented the "organs of generation, or principle of life."—John H. Thomson. New Bedford, Massachusetts, 1882. 68138. Gift of John H. Thomson.

ABORIGINAL APPARATUS.

IMPLEMENTS USED BY THE INDIANS OF CAPE FLATTERY AND THE ESKIMO TRIBES OF THE ARCTIC REGIONS.

THE INDIANS OF CAPE FLATTERY.

The Indians of Cape Flattery are the only representatives of their race south of Alaska who engage actively and energetically (for Indians) in whaling within the limits of the United States. It may, therefore, be of interest to give some account of this people; and to that end I have compiled the following data from the "Indians of Cape Flattery"* by James G. Swan:

The Makah Indians inhabit the region of Cape Flattery, at the entrance to the Strait of Fuca, Washington Territory, reserved for them under the "treaty of Neah Bay," in 1855. They are of medium height, with a good development of muscle, some of them being well proportioned and of unusual strength. Some have black hair, very dark brown eyes, and dark copper-colored skin; others have reddish hair, and a few have flaxen locks, light-brown eyes and fair skin, which may be attributed to an admixture of white blood of Spanish and Russian stock.† Their tribal name is "Kwe-nait-che-chat."

All matters pertaining to the government of this tribe are submitted to a council, at which the opinions of the old men usually prevail, though the women are permitted to speak on subjects pertaining to their rights or in which they are concerned. Formerly the strongest chief, possessed of the most friends and the greatest influence, governed the tribe, but at present, notwithstanding there are several in every village who claim to be descendants of chiefs, their power as rulers is not recognized, though they are treated as belonging to the aristocracy, and are listened to in council. They are also invited to the feasts when councils are held, receive a share of all presents, and their proportion of whales.

The Makahs are temperate, perhaps from a virtue of necessity, as the sale of intoxicating liquors is prohibited on the reservation. They are not active in vocations or pursuits other than fishing and whaling, and obtain some of their supplies by barter from neighboring tribes and white men. They devote very little time to agricultural pursuits or to the capture of land animals, but excel in the management of canoes, making long voyages from land for fish, and fearlessly attacking the whale. They manufacture their own fishing apparatus, and take es-

---
* Smithsonian Contributions to Knowledge, 220.
† "In Holmberg's work will be found an account of the wreck of a Russian ship, the survivors of whose crew lived several years among the Makahs. As late as 1854 I saw their descendants, who bore in their features unmistakable evidence of their origin."
—*George Gibbs.*

pecial pains with their harpoons and lances, for which instruments they have the greatest regard. The principal implements used by the Makah whalers are harpoons, lances, ropes, and buoys. The harpoon-heads were formerly made of shell, but at present are of sheet copper or steel, with barbs of elk or deer horn, tightly seized to the blades by cords or strips of bark, the whole being covered with spruce gum. The laniards attached to the harpoon are made of the sinews of the whale twisted into a rope and served with fibers of nettle. The lances are made of metal, with sockets for the ends of the poles. The poles for the harpoons and lances are heavy and unwieldy, but durable and strong. The buoys are made of seal skin with the hair inside, inflated when used and attached to the harpoon-laniards. These buoys are used for the double purpose of impeding the progress of the whale, so as to enable the Indians to kill it, and to prevent the animal from sinking when dead. The ropes used in towing whales ashore are made from the tapering limbs of the cedar and the long fibrous roots of the spruce. They are cut in lengths of three or four feet, and roasted or steamed in ashes, a process which renders them tough, pliable, and easy to split. They are then reduced to fine strands with knives, twisted, and made into ropes by being rolled between the palm of the hand and the naked thigh. All whaling implements that have been used in the capture are regarded with especial favor and handed down from generation to generation, and it is deemed unlucky to part with them. These Indians did not acquire the art of whaling from white men, and still employ the apparatus and processes which have come to them through countless generations. One point deserves especial consideration. The process of wrapping their harpoon-laniards, commonly known as "serving," has been in use by all sea-faring men for a number of years. The Makah Indian has his "serving-stick" and mallet, manufactures his twine from the fibers of the nettle, and "serves" his lines as neatly as do the fishermen of the eastern coast, and it is said they were familiar with the process before the advent of the whites.

The principal articles manufactured by the Makahs are canoes, whaling implements, conical hats, bark mats, fishing-lines, fish-hooks, knives, daggers, bows and arrows, dog-hair blankets, &c. Their largest and best canoes are made by the Clyoquots and Nittinats on Vancouver Island. Canoes of the medium and small sizes are made by the Makahs from cedar, procured a short distance up the Strait or on the Tseuss River. Before the introduction of iron tools the labor of making canoes was attended with many difficulties, the Indian hatchets being made of stone and the chisels of mussel shells ground to a sharp edge with pieces of sandstone. Naturally it required much time and labor to fell a large cedar, and it was only the wealthy chiefs, owning a number of slaves, that attempted such large operations. The tree was literally chipped away with their stone hatchets, or gnawed down after the fashion of beavers. After felling the tree many months were consumed in shaping the canoe. At present, however, they possess rude axes for rough hewing, and a peculiar form of chisel which may be used like a cooper's adze. Still, the process is very slow. The Indian is guided solely by the eye in modeling his canoe, and seldom, if ever, uses a measure of any kind, yet his lines are perfect and graceful. He also bends the wood, when necessary, by steaming it. The inside of the log is filled with water, which is heated with red-hot stones, a slow fire being made on the outside, near enough to warm the cedar without burning it. As the projections for the head and stern pieces cannot be cut from the same log, they are carved from separate pieces and "scarfed" by means of cedar withes held in their places by wooden pegs. The joints by this process are so perfectly matched as to be water-tight without calking. When the canoe is finished the interior is painted with a mixture of oil and red ocher. Sometimes charcoal and oil are rubbed on the outside, but more commonly it is simply charred, the surface being rubbed smooth with grass or cedar twigs. The paddles are made of yew, and are usually procured from the Clyoquots. The blade is broad, but tapers at the point. The paddles are also blackened by charring them in the fire, and afterwards polished. The sails were formerly made of mats of cedar bark, and such are still used by some of the Clyoquots, though some of the tribes in the vicinity now use cotton sails. The usual form is square, with yards at top and bottom, and the sail may be rapidly hoisted or lowered by means of a line which passes through a hole in the top of the mast. By rolling the sail around the lower yard it can be let out or shortened, as the occasion may require. Some of the Indians have adopted sprit-sails, but they are not in general use.

Blankets, which constitute the principal item of wealth, are made of feathers or down, of dog's hair, and of cedar bark. The manufacture of mats is the principal employment of the females during the winter, and for this purpose cedar bark is chiefly used. Baskets of various kinds are also made of this bark, but those intended for carrying heavy weights are made from spruce roots. Conical hats for the Indians are made of spruce roots split into fine fibers and plaited so as to be impervious to water, and painted of a black ground with red figures. The black is produced by grinding bituminous coal with salmon eggs, which are chewed up and spit on a stone. The hats sold to white men, however, resemble the common straw hat, and are made of spruce roots, some being of a plain buff color, while others have woven designs of various kinds. Recently they have commenced to cover bottles or vials with basket-work, for sale to seekers of Indian curiosities. Their fishing and whaling capes are made something like a "poncho," from cedar bark or from strips of cloth or old blankets. Their bows are usually made from yew, principally by the boys, and the arrows from split cedar. The arrow-heads are made of pieces of wire, bone, wood and bone combined, iron, or copper. The prongs of the bird-spears are made either of wood or bone, and the barbs of the fish-spears of iron or bone. The manufacture of whaling implements, particularly the harpoon poles and heads, is confined to individuals who dispose of them to the others. None of the Indians seem to have regular trades, yet the most expert confine themselves to certain branches. Some are skillful in working iron and copper, others in carving or painting, while others, again, are more expert in catching fish or killing whales.

The Indians do not understand the art of manufacturing pottery, although clay is found at Neah Bay. Their ancient utensils for boiling were simply wooden troughs, and the method of cooking in them was by hot stones. These troughs are used by many at present, especially on occasions of feasting, when a large quantity of food is necessary; but for ordinary purposes iron pots, brass kettles, and tin pans, which have been purchased from white traders, are used. Vessels for carrying water, and boxes for containing blankets or clothing, are made from boards, bent, when necessary, by the application of warm water; but these are manufactured principally by the Clyoquot Indians, very few being made by the Makahs. Wooden bowls and dishes, and chopping trays, are made from alder; but some of the bowls are made of knobs taken from decayed logs of maple or fir. Fishing-lines are made of kelp stems; halibut hooks from hemlock knots—whale sinew being used for tying on the bait. The barbs of the codfish hooks are made from bone, lashed to wooden shanks, for the capture of small fish, such as perch and rock. Small pieces of bone, sharp as needles at both ends, known as "gorge hooks," are seized in the middle by lines of sinew. The fish-club is usually a rough piece of wood, though sometimes rudely carved. In the manufacture of their tools the Makahs use a large stone for an anvil and a smaller one for a hammer. Their knives, which are employed either as weapons of defense or for cutting blubber or sticks, are made of rasps and files, the handles being made of bone and sometimes ornamented with brass or copper. The Makahs understand the art of tempering their knives. The chisels are made of rasps or any other kind of steel. The instruments for boring holes are simply pieces of iron or steel wire, flattened at the point and sharpened, with a rough stick as a handle. Cutting with a knife of any kind, or with a chisel, is done by working toward, instead of from, the person; but when they are so fortunate as to obtain an old plane they work it in the regular way. They also manufacture small knife-blades, which are inserted into wooden handles and used for whittling or scarifying their bodies during their medicine or "Ta-ma-na-was" performances. The common hammer is simply a stone; others used to drive wedges are manufactured with more care and in the form of a pestle.

Before the advent of the white man these Indians used dried halibut in place of bread, oil in place of butter, and blubber instead of beef or pork. When potatoes were introduced they formed a valuable addition to the food of the Indians; and since the white men have become more numerous the Indians have accustomed themselves to other articles of diet, such as flour, hard bread, rice, and beans, which are always acceptable to them. They are also fond of molasses and sugar, for which they are ever anxious to trade their furs, oil, or fish. Next in importance to the halibut are the salmon and a species of fish known as the "cultus," or bastard cod, which are usually eaten fresh except in seasons of great plenty, when the salmon are smoked. They capture all of the fish with the hook, using herring as bait. The squid is used as food and also as bait for halibut. Skates, though abundant, are seldom eaten, because they make their appearance during the halibut season. Three varieties of *Echinus* are abundant and eaten in great quantities. Mussels, barnacles, crabs, sea-slugs, perriwinkles, and limpets furnish occasional repasts. Scallops are excluded from their list of food, but their shells are used as rattles in ceremonials. Although oysters are found in the bays and inlets of Vancouver Island the Indians do not eat them.

Of land animals they eat the flesh of the elk, deer, and bear; but smaller animals, such as raccoons, squirrels, and rabbits, are seldom, if ever, eaten by them, and are killed only for their skins. They are particularly fond of sea-fowl, including pelicans, loons, cormorants, ducks of several kinds, grebes, and divers of various sorts. The roots of certain ferns, some species of meadow grass, water-plants, and several kinds of sea-weed, particularly eel-grass, are also used as food, as well as the young sprouts and fruit of the "salmon berry" and "thumb berry." Their method of serving up food is very primitive, the same forms being observed by all. The food is served in courses, and, when feasts are given, the guests are expected to carry away what they cannot eat. The host is offended if his guests do not partake of everthing that is placed before them, and if strangers are among the visitors it is not uncommon for four or five feasts to be given in the course of a single day or evening. An Indian is looked upon as a welcome guest who does justice to the hospitality of his host, and, in order that he may not offend any one, thrusts his fingers down his throat and throws off a load from his stomach to enable him to be prepared for the next feast. Although smoking is not universally practiced among them, they sometimes indulge in a whiff of tobacco mixed with dried leaves, after eating, fishing, and whaling. The "pipe of peace" is unknown among them.

Dog-fish are taken in large quantities for the oil contained in the liver, which forms the principal article of traffic between these Indians and the whites. But the fish itself is seldom eaten by the Makahs, unless the oil has been thoroughly removed. Dog-fish oil has a nauseous taste, and is not relished by these Indians, who are epicures in their way and prefer the oils of whales and seals. A very large species of shark, known among whalemen as "bone-shark," is occasionally killed by the Makahs on account of the great quantities of oil found in the liver. A fish of the genus *Anarrhichthys*, called the "doctor-fish," is only eaten by the medicine men. Porpoises are highly esteemed for food. Seals also abound. The skin of the hair-seal is taken off whole, blown full of air, and dried with the hair side in. It is used as a buoy in the capture of the whale, and is usually painted on the outside with rude devices in red, vermilion, or ocher. The seals, though sometimes killed with spears, are often shot with guns; but when they congregate during the breeding season, the Indians approach them with torches and clubs and kill numbers by knocking them on the head. The flesh of all the species of seal is eaten, and the skins of the fur-seals are sold to the whites.

The abundant supply of marine food, and the ease with which the Indians can obtain their subsistence from the ocean, makes them improvident in laying in supplies, with the exception of halibut, for winter use. On any day during the year, when the weather is favorable, they can procure provisions enough in a few hours to last them for several days.

The usual dress of the men consists of a shirt and blanket, the old men being content with the blanket only. Nearly all of them, however, have suits of clothing obtained from white persons, but these are only worn on arrival of strangers or when the Indians work for the whites, and they usually take them off at night, when they return to their lodges. During rainy weather they wear, in addition to blankets, conical hats and bear-skin cloaks. When whaling, they wear a bear-skin thrown over the shoulders; and when fishing, a small cape made from the fibers of bark. The women usually wear a shirt or long chemise, reaching from the neck to the feet; and some of them have, in addition, calico shirts tied as petticoats around their waists, or petticoats made of blankets or other coarse material. Formerly their dress was merely a blanket and a cincture of fringed bark reaching from the waist to the knees. The young women of the present day sometimes dress themselves in calico gowns or plaid shawls of bright colors. They also wear glass beads of various colors and sizes about their neck and ankles, with perhaps a dozen or more of bracelets made of brass wire around each wrist, nose and ear-ornaments composed of shells, beads, and strips of leather, and paint their faces with grease and vermilion. Both sexes wear nose-pendants, usually made from small pieces of *Haliotis* shell. The men wear their hair long; but when whaling they tie it up in a knot behind the head. They also decorate themselves by winding wreaths of evergreens around the knob of hair, or stick in sprigs of spruce and feathers. This head-dress is sometimes varied by substituting a wreath of sea-weed, or a bunch of cedar bark in the form of a turban. They paint their faces either black or red, or in stripes of various colors.

The Makahs claim that they were created on the Cape, and that animals were first produced. The first men sprang from an intimate intercourse of a star, which fell from heaven, with some of the animals; and from their offspring came the races of Nittinats, Clyoquots, and Makahs.

They believe that all living things—trees, birds, fishes, and animals—were formerly Indians, who, on account of their wickedness, were transformed into the shapes in which they now appear. They also believe that two men, "brothers of the sun and moon," and termed "Ho-hó-e-ap-béss," or the "men who change things," came on earth and made the transformation. The seal was a pilfering Indian, and therefore his arms were shortened and his legs tied so that he could only move his feet. He was cast into the sea and told to catch fish for his food. The mink was a great liar, and full of rascalities which he practiced on every one. The blue-jay was the mother of the mink. The raven was a strong Indian fond of flesh, and, in fact, a sort of cannibal; and the crow was his wife. The crane was a great fisherman. The king-fisher was also a fisherman, but a great thief. He stole a necklace made of shells, and this accounts for the ring of white feathers about his neck.

The Makahs, in common with all the coast tribes, hold slaves. In former times it is said the slaves were treated very harshly, and their lives were of no more value than those of dogs. The treaty between the United States and the Makahs makes it obligatory on this tribe to free their slaves, and although this provision has not thus far been enforced, it has had the effect of securing to the latter better treatment than they formerly had. Sometimes the master marries his slave woman, or a mistress takes her slave man as her husband; the offspring in such cases are regarded as half-slaves, and though some of them have acquired wealth and influence among the tribe, yet the fact that their fathers or mothers were slaves is considered as a stigma not to be removed for several generations. The slaves appear to have no task-work assigned them, but pursue the same avocations as their masters. The men assist in the fisheries, and the women manufacture mats and baskets or engage in domestic duties. Before the reservation was placed under the charge of an agent of the Government, it was considered degrading for a chief or the owner of slaves to perform any labor except to hunt, fish, or kill whales, but latterly no distinction is made between master and slave, but both are treated alike.

They keep little record of time, but have names for the different months, or "moons," twelve of which constitute two periods, the warm and cold. They remember and speak of a few days or of a few months, but of years, according to our computation, they know nothing. Their "year" consists of six "moons." The first of these periods commences in December, when the days begin to lengthen, and continues until June, when, as the sun recedes and the days shorten, another period commences and lasts until the shortest days. The seasons, however, are recognized by them as they are by ourselves, namely, spring, summer, autumn, and winter. The names of the months are as follows:

December is called the "moon" in which the chet-a-pook, or the California gray whale, makes its appearance.

January is the "moon" in which the whale has its young.

February, the "moon" when the weather begins to grow better and the days are longer, and when the women begin to venture out in canoes after firewood without the men.

March, the "moon" when the finback whales arrive.

April, the "moon" of sprouts and buds.

May, the "moon" of the strawberry and "salmon-berry."

June, the "moon" of the red huckleberry.

July, the "moon" of wild currants, gooseberries, etc.

August is a season of rest. No fish are taken or berries picked, except occasionally by children or idle persons.

September, work of all kinds commences, particularly cutting wood, splitting boards, and making canoes.

October is the "moon" for catching the "tsa-tar-wha," a variety of rock-fish, by means of a trolling-line, with a bladder buoy at each end and a number of hooks attached.

November is the season of winds and screaming birds.

The winds are the breath of fabulous beings who reside in the quarters whence they come, representing the different points of the compass. The Indians are excellent judges of the weather and can predict a storm or calm with almost the accuracy of a barometer.

Both males and females are passionately fond of gambling, and continue their games for days at a time, or until one party or the other loses all it has. They have several kinds of gambling instruments; and one game in particular, common to all the Indians of this Territory, and called in their jargon "la-hull," is played with disks made of hazel-wood, conclusions being arrived at by guessing, as is the case in the majority of their games. Another game consists in passing a stick rapidly from hand to hand, the object being to guess in which hand it may be. A third game is played by females, with four beaver teeth, marked on one side and plain on the other, which are thrown like dice.

When a Makah dies the body is immediately rolled in blankets and firmly bound with ropes and cords, then doubled up in the smallest compass and placed in a box, which is also firmly bound with a rope. A portion of the roof is removed, and the box with the body is taken out at the top of the house and lowered to the ground, from a superstition that if a dead body is carried through a door-way any person who passes through it afterwards will immediately sicken and die. It was formerly the custom to deposit the body in a tree, but of late years it has been buried in the earth, with a portion of the property of the deceased placed on top of the box. If a man, his fishing or whaling-gear, gun with lock removed, or his clothing and bedding are buried with him. If a woman, her beads, bracelets, calico garments and other wearing apparel, and baskets are buried with her. A little earth is thrown on the box and property, and the space filled in with stones. The grave is then decorated with blankets, calico shawls, handkerchiefs, looking-glasses, crockery, tin-ware, and implements used in digging the grave.

No particular order is observed in the arrangement of these articles, but they are usually placed according to the fancy of the relatives of the deceased.

Several varieties of the whale are taken at different seasons, some being captured, and others, including the right whale, drift ashore, having been killed by whalemen, sword-fish, or other agencies. The California gray whale is the kind usually captured by the Makahs, the others being rarely attacked. Among the various species of whales found off this coast may be mentioned the sperm-whale, which is rarely seen, the right whale, sulphur-bottom, finback, blackfish, killer, and as just referred to, the California gray whale.

As the method of whaling peculiar to these Indians forms the most important topic in connection with this paper, I quote herewith at length from Mr. Swan. He says:

"Their method of whaling, being both novel and interesting, will require a minute description—not only the implements used, but the mode of attack and the final disposition of the whale being entirely different from the practice of our own whalemen. The harpoon consists of a barbed head, to which is attached a rope or lanyard, always of the same length, about 5 fathoms, or 30 feet. This lanyard is made of whale's sinews twisted into a rope about an inch and a half in circumference, and covered with twine wound around it very tightly, called by sailors "serving." The rope is exceedingly strong and very pliable.

"The harpoon-head is a flat piece of iron or copper, usually a saw-blade or a piece of sheet copper, to which a couple of barbs made of elk's or deer's horn are secured, and the whole covered with a coating of spruce gum. The staff is made of yew in two pieces, which are joined in the middle by a very neat scarph, firmly secured by a narrow strip of bark wound around it very tightly. I do not know why these staves or handles are not made of one piece; it may be that the yew does not grow sufficiently straight to afford the required length; but I have never seen a staff that was not constructed as here described. The length is eighteen feet; thickest in the center, where it is joined together, and tapering thence to both ends. To be used, the staff is inserted into the barbed head and the end of the lanyard made fast to a buoy, which is simply a seal-skin taken from the animal whole, the hair being left inwards. The apertures of the head, feet, and tail are tied up air-tight and the skin inflated like a bladder.

"When the harpoon is driven into a whale the barb and buoy remain fastened to him, but the staff comes out, and is taken into the canoe The harpoon which is thrown into the head of the whale has but one buoy attached, but those thrown into the body have as many as can be conveniently tied on; and, when a number of canoes join in the attack, it is not unusual for from thirty to forty of these buoys to be made fast to the whale, which, of course, cannot sink, and is easily dispatched by their spears and lances. The buoys are fastened together by means of a stout line made of spruce roots, first slightly roasted in hot ashes, then split with knives into fine fibers, and finally twisted into ropes, which are very strong and durable. These ropes are also used for towing the dead whale to the shore. The harpoon-head is called kwe-paptl; the barbs, tsa-kwat; the blade, kūt-só-wit; the lanyard attached to the head, klūks-ko; the loop at the end of the lanyard, kle-tait-lĭsh; the staff of the harpoon, du-pói-ak; the buoy, dōpt-kó-kuptl; and the buoy-rope, tsis-ka-pūb.

"A whaling canoe invariably carries eight men: one in the bow, who is the harpooner, one in the stern to steer, and six to paddle. The canoe is divided by sticks, which serve as stretchers or thwarts, into six spaces, named as follows: The bow, he-tuk-wad; the space immediately behind, ka-kai-woks; center of canoe, cha-t'-hluk-dōs; next space, he-stuk'-stas; stern, kli-chá. This canoe is called pa-dan-t'-hl. A canoe that carries six persons, or one of medium size, is called bo-kwis'-tat; a smaller size, a-tlis-tat; and very small ones for fishing, te-ka-aú-da.

"When whales are in sight, and one or more canoes have put off in pursuit, it is usual for some one to be on the look-out from a high position, so that in case a whale is struck a signal can be given and other canoes go to assist. When the whale is dead it is towed ashore to the most convenient spot, if possible to one of the villages, and hauled as high on the beach as it can be floated. As soon as the tide recedes, all hands swarm around the carcass with their knives, and in a very short time the blubber is stripped off in blocks about two feet square. The portion of blubber forming a saddle, taken from between the head and dorsal fin, is esteemed the most choice, and is always the property of the person who first strikes the whale. The other portions are distributed according to rule, each man knowing what he is to receive. The saddle is termed u-butsk. It is placed across a pole supported by two stout posts. At each end of the pole are hung the harpoons and lines with which the whale was killed. Next to the blubber at each end are the whale's eyes; eagle's feathers are stuck in a row along the top, a bunch of feathers at each end, and the whole covered over with spots and patches of down. Underneath the blubber is a trough to catch the oil which drips out. The u-butsk remains in a conspicuous part of the lodge until it is considered ripe enough to eat, when a feast is held, and the whole devoured or carried off by the guests, who are at liberty to carry away what they cannot eat. After the blubber is removed into the lodge the black skin is first taken off, and either eaten raw or else boiled. It looks like India rubber; but though very repulsive to the eye it is by no means unpalatable, and is usually given to the children, who are very fond of it, and manage to besmear their faces with the grease till they are in a filthy condition.

"The blubber, after being skinned, is cut into strips and boiled, to get out the oil that can be extracted by that process; this oil is carefully skimmed from the pots with clam-shells. The blubber is then hung in the smoke to dry, and when cured looks very much like citron. It is somewhat tougher than pork, but sweet (if the whale has been recently killed), and has none of that nauseous taste which the whites attribute to it. When cooked it is common to boil the strips about twenty minutes, but it is often eaten cold and as an accompaniment to dried halibut.

"From information I obtained I infer that formerly the Indians were more successful in killing whales than they have been of late years. Whether the whales were more numerous, or that the Indians, being now able to procure other food from the whites, have become indifferent to the pursuit, I cannot say; but I have not noticed any marked activity among them, and when they do go out they rarely take a prize. They are more successful in their whaling in some seasons than in others, and whenever a surplus of oil or blubber is on hand it is exchanged or traded with Indians of other tribes, who appear quite as fond of the luxury as the Makahs. The oil sold by these whalers to the white traders is dog-fish oil, which is not eaten by this tribe, although the Clyoquot and Nootkan Indians use it with their food. There is no portion of a whale, except the vertebræ and offal, which is useless to the Indians. The blubber and flesh serve for food; the sinews are prepared and made into ropes, cords, and bowstrings, and the stomach and intestines are carefully sorted and inflated, and, when dried, are used to hold oil. Whale oil serves the same purpose with these Indians that butter does with civilized people; they dip their dried halibut into it while eating, and use it with bread, potatoes, and various kind of berries. When fresh it is by no means unpalatable; and it is only after being badly boiled, or by long exposure, that it becomes rancid, and as offensive to a white man's palate as the common lamp oil of the shops."

MAKAH INDIANS.

WHALING IMPLEMENTS EMPLOYED BY THE INDIANS OF CAPE FLATTERY, COLLECTED BY JAMES G. SWAN.

[Compiled from explanatory notes accompanying the objects.]

CAPTURING THE WHALE.

HARPOONS.

HEADS AND LANIARDS.

HARPOON HEAD AND LANIARD.

Head, apparently a piece of an old saw blade, covered with a coating of spruce gum. Laniard, sinews of the whale served with twine made from fibers of nettle to render it impermeable to water. Barbs, elk bone; sheath, bark. Length, 20 feet. Makah Indians, Cape Flattery, 1883. 72635. James G. Swan. Used by natives for fastening seal-skin buoys to whales.

HARPOON HEAD AND LANIARD.

Head made of piece of sheet-brass; barbs, elk-bone, ornamented, covered with a coating of spruce gum. Laniard, sinews of the whale neatly *laid up*, and served with twine to keep out water, which is injurious to the fibers. Sheath, bark. Makah Indians, Cape Flattery, 1883. 72634. James G. Swan. The harpoons formerly used by these Indians were made of mussel shells; at present of copper sheathing, brass, or old saw-blades. The serving for the laniards was formerly made exclusively from the fibers of the nettle, which are also used now by the old men; and though the young men, in some instances, use cotton twine, yet they prefer the nettle. A harpoon that has been successfully used acquires additional value.

HARPOON AND LANIARD.

Harpoon and line attached to pole and seal-skin buoy, showing the manner in which the apparatus is rigged when used. Headpiece of sheet brass. Laniard, whale-sinew, served with twine made from the fibers of the nettle. Makah Indians, Cape Flattery. 72752. James G. Swan. The harpoon is not permanently fastened to the staff; it is, however, connected with the buoy by means of a laniard. When the harpoon is thrust into the whale, the staff is withdrawn and taken into the canoe, and the animal is incumbered with the buoy. A harpoon with one buoy attached is thrown into the head of the whale, but the harpoon thrown into the body has as many buoys as can

HARPOON AND LANIARD—Continued.
conveniently be tied on; and, when a number of canoes join in the attack, it is not unusual for from thirty to forty of these floats to be made fast to one whale, which, of course, cannot sink, and is easily dispatched by the spears and lances. The Indians did not acquire the art of whaling from white men; it has been handed down through countless generations. The same kind of apparatus has also been in use for many years.

HARPOON-POLES.

HARPOON-POLE.
A heavy, unwieldy pole made of yew (*Taxus brevifolia*), scarfed in three places, and served with strips of wild-cherry bark. One end tapers to a point for the reception of harpoon-socket. Used by natives in thrusting the harpoon into the whale to make fast the seal-skin buoys. Length, 15 feet. Makah Indians, Cape Flattery, 1883. 26825. James G. Swan. An implement for which the Makah whaler has a special regard. It is seldom used without being broken; it is then repaired, and acquires additional value. I saw one with six places where it had been repaired, and the owner would not part with it for any price. It was difficult to get the one now sent, although they were perfectly willing to make me new ones, but had no yew. Some of these harpoon staffs which have been in the same family for many generations could not be purchased, from a superstition that it would be unlucky.

IMPEDING THE PROGRESS OF THE WHALE.

FLOATS.

SEAL-SKIN BUOY.
Skin of the hair-seal used by natives in the capture of the whale. Indian name, "Do-ko-kuptl." New. Length, 36 inches. Makah Indians, Cape Flattery, 1883. 72629. James G. Swan.

SEAL-SKIN BUOY.
Skin of hair-seal, small stationary wooden toggle at either end for holding eye-splice of harpoon-line. Small laniards made of fibers of spruce roots, for *making fast* to other buoys. Indian name, "Do-ko-kuptl." Length, 38 inches. Makah Indians, Cape Flattery, 1883. 72630. James G. Swan. Inflated and attached to the harpoon, showing the manner in which the apparatus is used during the capture. A number of buoys being made fast to the whale prevents its progressive motions, thus affording the natives an opportunity to kill it with the lance (72674).

FISHERIES OF THE UNITED STATES.

SEAL-SKIN BUOY.
Skin of hair-seal used by natives in the capture of the whale. Indian name, "Do-ko-kuptl." Old. Length, 34 inches. Makah Indians, Cape Flattery, 18—. James G. Swan.

KILLING THE WHALE.

LANCE-HEADS.

LANCE-HEAD.
New. Indian name, "Kathlat-te-uk." Head, steel; socket, wood, served with bark strips. Covered with a coating of spruce gum. Length, 7 inches. Makah Indians, Cape Flattery, 1883. 72639. James G. Swan. Used with a long pole (72674), and when thrust into a whale the lance becomes detached, and is recovered when the whale is cut up. A lance-head that has been successfully used acquires additional value, and for some of them the Indians ask a fabulous price.

LANCE-HEAD.
An old lance-head formerly the property of Haiks, at one time a chief of the Neah Bay. He made it many years ago from a piece of a musket-barrel. It was highly prized by the relatives. An ingenious and simple device. Piece of gun-barrel hammered into the shape of a lanceolate blade, the rear portion of barrel serving as a socket. Indian name "Kathlat-te-uk." Length, 7 inches. Makah Indians, Cape Flattery, 1882. 72640. James G. Swan. Attached to lance-pole and used in killing whales. Thrust into the most vulnerable parts of the whale; the pole is withdrawn, and the head regained when the whale is cut up. Lances that have been used are greatly enhanced in value.

LANCE-POLES.

LANCE-POLE.
Long, heavy, and unwieldy pole, with separate pieces serving as *shanks* seized to either end. Lance-head attached. The form of this staff, with its long, tapered point, is to enable the Indian to thrust it as deeply as possible into the most vulnerable parts of the whale. After a sufficient number of skin-buoys have been fastened to the whale to prevent it from remaining under water, and when it is nearly exhausted from the harpoons which have been thrust into it, an Indian places himself in the bow of the canoe with his face towards the stern; the canoe is then paddled alongside the whale, and, standing up with one foot on the thwart and the other on the gunwale of the canoe, the Indian raises the staff high above his head and thrusts the lance as deep into the whale as he can, using his utmost force. The heart is the place aimed for, and, if successful, the lance-

LANCE-POLE—Continued.

head being detached remains, and the animal dies at once. If a vital portion is not struck at first, other lance-heads are thrust in until the death wound is given. The whale is towed ashore, cut up, and the lance-head secured. Length, 20 feet 4 inches. Makah Indians, Cape Flattery, 1883. 72674. James G. Swan.

TOWING WHALES ASHORE.

TOW-LINES.

TOW-LINE.

Small tow-line, "ses-tope," made of fibers of spruce roots. Used by natives for towing the whale ashore. Port Townsend, Washington Territory. 72633. James G. Swan. Makah Indians, Cape Flattery.

TOW-LINE.

Small size. Indian name "ses-tope." Made of spruce roots (*Abies Douglasii*). The process of manufacture consists in (1) roasting the material in hot ashes; (2) splitting with knives into fine fibers; and (3) twisting the fibers into a rope. Durable and strong. Makah Indians, Cape Flattery, 1883. 72631. James G. Swan. Used by natives in towing whales ashore.

TOW-LINE.

New. Large size. Made of fibers of spruce roots (*Abies Douglasii*). The long slender roots are first roasted in the ashes, then split into fine strands with knives, twisted, and laid up into ropes by hand. These ropes are beautifully made, exceedingly strong, and buoyant. The Indians not only understand the art of rope-making by hand as well as the whites, but they can also *knot* and *graft* a rope as well as white sailors. Makah Indians, Cape Flattery, 1883. 72632. James G. Swan. Used by natives for towing whales ashore.

PADDLES.

WHALING PADDLE.

Made of yew; the common form adopted by the natives in whaling. The paddle has a long, tapering point to enable the canoe to silently approach a whale, as the blade can be thrust deep in the water and the reverse stroke made with comparatively little splashing or noise. Length, 5 feet. Makah Indians, Cape Flattery, Washington Territory. 72676. James G. Swan.

WHALEMAN'S CLOTHING.

BEAR-SKIN CLOAK.

Indian name, "Artleitquitl." Worn by natives when whaling or fishing, or in wet weather on shore. 74 by 43 inches. Makah Indians, Cape Flattery, 1883. 72693. James G. Swan.

PRODUCTS OF THE WHALE.

BALEEN.
Nine slabs of whalebone from the sulphur bottom whale. Makah Indians, Neah Bay, Washington Territory. 72692. James G. Swan.

IMPLEMENTS USED IN THE CAPTURE OF THE SEAL.

SPEARS, HEADS, AND LANYARDS.

SEAL SPEAR.
A slender staff or pole, with two prongs of unequal lengths upon which are placed respectively two metal heads with one barb each. The spear-heads are held in place by laniards which are hauled taut and firmly grasped with the pole in the left hand. When used the ends of the laniards are attached to a long line, one end of which remains in the boat. The butt of the pole is provided with a flaring piece of wood which is used as a finger-rest when the Indian thrusts the instrument into the seal. Length, 15 feet 10 inches. Makah Indians, Neah Bay, Washington Territory. 72671. James G. Swan. Used by the natives in killing fur seals. The canoe is paddled silently to a short distance from the sleeping seal, and the spear thrust forcibly into the animal. The canoe is hauled by means of the rope closer to the seal, which is dispatched by a blow on the head with a club. The Indians invariably smash in the skull of a seal even when the animal is killed by the thrust of the spear, which is frequently the case. So universal is this practice that I was unable, during a residence of three years at Neah Bay, to obtain a perfect specimen of a skull, although hundreds of skulls may be seen on the beach any day during the sealing season, but every one was fractured. [JAMES G. SWAN.]

STAFF FOR SEAL SPEAR.
Slender pole with two prongs, without spears, and finger-rest at rear end. Used for killing seal. Length, ———. Makah Indians, Cape Flattery, 1883. 72670. James G. Swan.

RECEPTACLES FOR SEALING IMPLEMENTS.

BASKET.
A large basket, "Kla-ash," used by natives for holding spear-heads, harpoons, and lines, when sealing. Length, 28 inches. Port Townsend, Washington Territory, January, 1883. 72665. James G. Swan. These baskets are never offered for sale. The prices asked for them, when a native is induced to sell, exceed those for the ordinary baskets.

BASKET.
A small basket, "Kla-ash," used as a receptacle for spear-heads by natives when engaged in killing seal. Length, 15½ inches. Makah Indians, January, 1883. 72664. James G. Swan.

BASKET.
Basket used to hold spear-heads and other small articles when sealing—called by the Makahs, "Kla-ash." A very fine specimen. Double, made for a chief, and was procured as a special favor. Such baskets are never offered for sale. After having been used they acquire additional value, and to sell one is deemed unlucky. This being new, was more easily obtained. Length, 19 inches. Makah Indians, Cape Flattery, 1883. James G. Swan.

ACCESSORIES.

SERVING LANIARDS.

SERVING STICK AND TWINE.
Stick, yew; twine, nettle fiber. Ends of stick carved to represent the caudal fin of the whale. Used in connection with the mallet (76638) to serve harpoon laniards. Length, 16½ inches. Makah Indians, Cape Flattery, 1882. 72637. James G. Swan. By means of this implement and the mallet, twine is wound or wrapped around the harpoon lines in spiral folds in the same manner as ordinary seamen *serve* a rope with spun-yarn or marline. The Indians employed this process before the advent of the white man. The necessities of the case caused them to adopt a plan at once simple and effective. "This *stick* has been in the family from which it was procured more than four generations. It was the property of Chief Haiks, who died at Neah Bay thirty years ago. His whaling implements have been carefully preserved and never used since his death."—[J. G. SWAN.]

SERVING MALLET.
Indian name, "Kla-ta-bup." Small wooden mallet, square ends, longitudinal groove in upper surface; used with the *serving stick* (72637) in wrapping the sinew rope, for harpoon laniards, with twine; usually made from the fibers of nettle. Length, 6 inches. Makah Indians, Cape Flattery, 1882. 72638. James G. Swan.

MANUFACTURE OF TWINE.

BARK.
Inner bark of white cypress (*Cupressus nukatensis*), from which is manufactured the twine used in whaling, as well as soft beds for infants. Small package; length, 5 inches. Makah Indians, Cape Flattery. 72641. James G. Swan. When a harpoon

FISHERIES OF THE UNITED STATES.

BARK—Continued.

with one buoy attached has been darted into a whale, another buoy is immediately attached to the laniard of the first, the operation being repeated until a sufficient number of floats have been bent on. It is also often necessary to detach some of the buoys to make them fast to other harpoons and buoys. The twine made from cypress bark is especially well adapted for this purpose, as it breaks easily when wet, and quickly releases the buoys, which would not be the case with other kinds of twine.

IMPLEMENTS USED BY ESKIMO TRIBES IN THE CAPTURE OF THE SEAL, WALRUS, AND WHALE.

ESKIMO HARPOONS.

ESKIMO WHALING-HARPOON.

Pole, wood; ivory tip recessed for walrus ivory spear or shank, which is lashed to the pole with thongs of raw hide. Length, 8 feet 3 inches. Northeast coast. 10265. Smithsonian Institution.

SEAL-HARPOON.

Eskimos. Igloolik. 10400. Captain C. F. Hall.

ESKIMO HARPOON.

Pole, wood; iron spur and ferule at butt; shank, iron; harpoon, ivory (lily iron); point tipped with iron; recessed for end of shank; rigid eye for line; line, seal-skin; sheath, wood. Length, 8 feet 5½ inches. Eskimos, Greenland. 19518. George Y. Nickerson. An improved whaling-harpoon made by natives, evidently, in part, from material obtained from a whaling-vessel.

WHALING-HARPOON.

Pole, wood; tipped with ivory. Ivory point fitting in recessed tip of pole, and seized to the handle with thongs of walrus-hide. Length, 8 feet. Eskimos, Greenland. 19519. George Y. Nickerson.

ESKIMO WHALING-HARPOON.

Pole, wood; butt-piece, ivory; inserted in a recess in the butt, and riveted with native copper. Grip, ivory; pole tipped with ivory, recessed for a spear or shank, 18½ inches long, made of walrus ivory, and lashed to pole with raw hide. Length, 8 feet 7 inches. Cumberland Gulf. 30008. W. A. Mintzer, U. S. N. May be used in the capture of whale or walrus.

ESKIMO SEAL-HARPOON.

Pole, wood, one inch in diameter; butt recessed to receive a recurved bone spear, which is lashed with seal-skin; ivory peg for grip, lashed to pole with seal-skin; tip mounted with a bulb-like ivory head recessed for shank; shank, ivory, fastened to line with a small seal-skin laniard or becket. Lily-iron, ivory, tipped with iron, rigid eye for line; seal-skin line attached to head. Total length, 9 feet 2 inches. Norton Sound, Alaska. 33888. E. W. Nelson. Combined harpoon and lance, manufactured and used by natives in the capture of seal.

SEAL-HARPOON.

Pole, wood; ivory spear or lance seized to butt with seal-skin; ivory grip; head-piece and shank, walrus ivory; tip of pole served with seal sinews. Harpoon wanting. Length, 9 feet 6 inches. Port Clarence, Alaska. 43429. E. W. Nelson. Harpoon and lance, or spear, combined.

SEAL-HARPOON.

Pole, wood; lance, walrus ivory, seized to butt with seal-thongs; Grip, ivory; carved in imitation of head of seal; tip of pole served with alternate layers of black and transparent strips of baleen; head-piece ivory recessed for ivory shank, lashed to pole with a seal-thong. Harpoon wanting. Length, 11 feet. Golorna Bay, Alaska. 43346. E. W. Nelson. Harpoon and lance combined.

ESKIMO HARPOON.

Handle, wood, tipped with head of an animal carved in bone. Bone shank inserted in recess of tip and lashed with raw-hide. End of handle near tip served with strips of baleen and raw-hide. Small seal-head carved in bone and seized to central part of handle, used as a finger rest, and as a stop for the line. Harpoon-butt and head wanting. Total length, 54 inches. Sledge Island. 45415. E. W. Nelson.

ESKIMO HARPOON-HANDLE.

Handle, wood, tipped with bone. Shank, bone, inserted in recessed head of tip, and lashed to handle with hide. Pole in two sections to fit case. Total length, 76 inches. Cape Lisburne, Alaska. 46177. W. H. Dall.

BELUGA-HARPOON OR WHALING STICK.

A light stick half an inch in diameter, with a walrus-ivory tip, carved in the shape of the head of an animal. A wooden plug is inserted in the mouth and recessed for the insertion of the neck or shank. Harpoon, bone tipped with slate. When the beluga is struck the head becomes detached from the shaft. Used in connection with the Throwing Stick. Length, 5 feet. Alaska. 72391. C. L. McKay.

BELUGA-HARPOON SHAFT.
Light wooden stick one-half inch in diameter, tipped with walrus-ivory, carved in shape of a head of an animal. Harpoon wanting. Length, 4 feet 5½ inches. Alaska. 72392. C. L. McKay. Used by natives in connection with the accompanying Throwing Stick (72398) for the capture of the beluga.

BELUGA-HARPOON SHAFT.
Wood, one half inch in diameter, with ornamental head carved in walrus-ivory. Harpoon wanting. Length, 4 feet 4 inches. Alaska. 72393. C. L. McKay. Used by natives in connection with the accompanying Throwing Stick (72398) in the capture of the beluga.

ESKIMO HARPOON-HEADS.

LILY-IRONS.

SEAL-HARPOON HEAD.
Walrus-ivory, with brass tip riveted with native copper. Laniard, raw-hide. Bights in the ends of laniard seized with seal sinew. Length of head, 3 inches. Length of laniard, 40 inches. Cape Lisburne, Arctic Ocean. (3627.) W. H. Dall. Old. Has been used.

HARPOON-HEAD.
Detachable harpoon-head, bone, with iron tip. Length, 6 inches. Eskimo, Cape Espenberg, Kotzebue Sound. (3700.) T. H. Bean.

THREE HARPOON-HEADS OF BONE AND IRON.
Recessed in rear end for poles. Rigid eyes for laniards. Length, 4⅞, 5⅜, 5⅝ inches. Eskimos, Northeast coast. 9838. S. F. Baird.

WALRUS-HARPOON HEAD.
End recessed for pole. Rigid eye for laniard. Length, 5 inches. Eskimos, Igloolik. 10136. Captain C. F. Hall.

PART OF ANCIENT INNUIT HARPOON-HEAD.
Bone. Length, 3¼ inches. Repulse Bay. 10404. Captain C. F. Hall.

THREE HARPOON-HEADS.
Ivory, tipped with iron. Recessed for poles and eyes for laniards. Tip of one broken. Length, 4½, 4¾, 5¼ inches. Eskimos, Cumberland Gulf. 29974. W. A. Mintzer, U. S. N.

HARPOON-HEAD OF BONE AND IRON.
Recessed for pole, rigid eye for laniard. Length, 5¼ inches. Eskimos, Cumberland Gulf. 29975. W. A. Mintzer, U. S. N.

SEAL-HARPOON HEAD.
Walrus-ivory, tipped with brass. New. Laniard, raw-hide. Bight seized with seal-sinew. Length of head, 3 inches; length of laniard, 59 inches. Eskimo, Cape Lisburne, Arctic Ocean. 46032. W. H. Dall.

SEAL-HARPOON HEAD.
　Ivory, tipped with brass. Laniard, raw-hide, seized with seal-sinews. Length of head, 3½ inches; length of laniard, 15 inches. Eskimo, Cape Lisburne, Alaska. 46033. W. H. Dall.

SEAL-HARPOON HEAD.
　Walrus-ivory, tipped with iron. Laniard, raw-hide. Bights seized with seal-sinew. Length of head, 3 inches; length of laniard, 51 inches. Eskimo, Cape Lisburne, Alaska. 46035. W. H. Dall.

SEAL-HARPOON HEAD.
　Walrus-ivory, with brass tip. Laniard, raw-hide. Seized with sinew of seal. Length of head, 3⅞ inches; length of laniard, 16¾ inches. Eskimo, Cape Lisburne, Alaska. 46038. W. H. Dall.

HARPOON-HEAD.
　A detachable harpoon-head made of bone and iron. Length, 5⅞ inches. Eskimo, Big Diomede Island. 46373. T. H. Bean.

HARPOON-HEAD.
　Detachable head, bone, with iron tip. Rear part of head bifurcated. Strap, raw-hide. Wooden neck or shank inserted in recess of head and fastened to strap with a small raw-hide becket. Length, 5½ inches. Nunivak, Alaska. 48240. E. W. Nelson.

HARPOON-HEAD.
　Detachable head, tipped with iron. Rear end bifurcated. Rawhide strap. Length of head, 5½ inches. Nunivak, Alaska. 48241. E. W. Nelson.

SEAL-HARPOON HEAD.
　Walrus-ivory, tipped with brass. Laniard, seal-hide. Length of head, 3 inches; Length of laniard, 8 feet 8 inches. Eskimo, Cape Lisburne, Alaska. 49034. W. H. Dall.

HARPOON-HEAD.
　A detachable harpoon-head, bone, tipped with brass. Length, 9 inches. Ooglaamie, 1882. 56601. Lieutenant P. H. Ray, U. S. A.

HARPOON-HEAD.
　A detachable harpoon-head, bone, with brass tip. Used by natives in capturing the whale. Length, 10¼ inches. Ooglaamie, 1882. 56602. Lieutenant P. H. Ray, U. S. A.

HARPOON-HEAD.
　A detachable bone harpoon-head, tipped with brass; recessed for end of pole; eye for strap. Strap, raw-hide. Length of head, 4¾ inches; length of strap, 12 inches. Ooglaamie, 1882. 56623. Lieutenant P. H. Ray, U. S. A.

HARPOON-HEAD.
Bone, tipped with iron. Used in beluga fishing with the accompanying harpoon-shafts (72392–93). Length, 9¼ inches. Alaska. 72394. C. L. McKay.

BONE HARPOON-HEAD.
Length, 7½ inches. Alaska. 72395. C. L. McKay.

ESKIMO LANCES—SEAL AND WALRUS.

SEAL-LANCE.
Pole, wood; lance-head, flint, lashed to pole and served with seal sinew; grip, ivory. Length, 10 feet 2 inches. Norton Sound. 33889. E. W. Nelson.

SEAL-LANCE.
Pole, wood, served at tip with strips of baleen; shank, ivory, seized with thong of seal-skin; lance-head, iron, riveted to shank. Length, 12 feet 2 inches. Poonook, Alaska. 15954. H. W. Elliott. Used by natives for the capture of seal and walrus.

ESKIMO LANCE.
Pole, wood, butt-piece seized to pole with seal sinew; grip, ivory; lance-blade, section of walrus tusk, twenty inches long, seized to pole and served with seal sinew. Length, 8 feet 3 inches. Alaska. 36062. E. W. Nelson. Used by natives to kill both seal and walrus.

ESKIMO LANCE.
Handle, wood, one and one-half inches in diameter; butt-piece, ivory, wedge-shaped, inserted in scarf in the butt of pole, and lashed and served with the sinew of the seal; ivory peg near tip of handle used as a finger-grip when manipulating the instrument; lance-blade, longitudinal section of walrus tusk, lashed to pole with seal-thong. Total length, 8 feet. Alaska. 36063. E. W. Nelson. Used by natives in the capture of seal and walrus.

ESKIMO LANCE.
Pole, wood, butt-piece, ivory, wedge shaped, inserted in scarf at butt of pole and served with seal sinew; grip, ivory; lance, walrus ivory, 18¾ inches long, lashed to pole and served with seal sinew. Length, 7 feet 6 inches. 43379. E. W. Nelson. Used by natives to kill both seal and walrus.

ESKIMO SEAL-LANCE.
Pole, wood; butt, ivory with wedge-shaped scarf for lance or spear; lance lashed to butt with seal-skin thong; finger-grip, ivory; tip of pole served with black and horn-colored baleen strips; head-piece and shank, walrus ivory; harpoon wanting. Length, 9 feet. Sledge Island, Alaska. 45416. E. W. Nelson. Harpoon and lance or spear combined.

SEAL-LANCE.
  Pole, wood; lance-head, flint, lashed to pole and seized with seal-sinew. Length, 9 feet. Cape Nome, Alaska. 45419. E. W. Nelson.

SEAL-LANCE AND HARPOON.
  Handle, wood; lance, walrus ivory, lashed to butt with seal-skin; butt and tip of pole served with strips of wood; head-piece, walrus ivory, recessed for harpoon shank, and lashed to pole with a thong of seal-skin; grip, ivory; harpoon wanting. Length, 12 feet. Eskimo, Cape Lisburne, Arctic Ocean. 46176. W. H. Dall. Lance and harpoon combined.

ESKIMO LANCE.
  Pole, wood; butt-piece, ivory, wedge-shaped, seized and served with seal sinew; grip, ivory, lashed to pole with seal sinew; tip of pole served with seal sinew, recessed for lance; lance, bone, 22 inches long, lashed to pole with thongs of seal-skin. Length, 8 feet. Nunivak Island, Alaska. 48377. E. W. Nelson. Used by natives to kill both seal and walrus.

ESKIMO LANCE.
  Pole, wood; butt-piece, ivory, served with seal sinew; rigid ivory grip; lance, piece of walrus tusk, seized to pole with seal sinew. Length, 7 feet 8 inches. Nunivak Island, Alaska. 48378. E. W. Nelson. Used by natives to kill both seal and walrus.

ESKIMO LANCE.
  Pole, wood; butt-piece seized to pole with seal sinew; grip, ivory; lance-blade, section of walrus tusk, 19 inches long, seized to pole, and served with seal sinew. Length, 8 feet 3 inches. Alaska. 48380. E. W. Nelson. Made and used by natives to kill both seal and walrus.

WALRUS-LANCE.
  Pole, wood; lance-head, flint, 4⅛ inches by 5 inches, inserted in recessed tip, lashed and served with seal sinew; pole in two sections to fit case. Total length, 20 feet 4 inches. Point Barrow, 1882. 56765. Lieutenant P. H. Ray, U. S. A.

SEAL-LANCE.
  A stout wooden handle, with walrus-ivory lance, hollowed on one side, and an ivory butt-piece. The lance is lashed to the handle with a seizing of gut, and further secured by a string from the inner side of tip. An ivory peg is fastened to the butt of the point or blade, by means of which the operator is assisted in steadying the lance when manipulating it. Length, 5 feet. Alaska. 72401. C. L. McKay.

BELUGA-LANCE BUTTS.
Two butts made of walrus ivory, wedge-shaped, so as to be conveniently driven into the end of the lance, and provided with shoulders, by means of which they are seized and lashed. Length, 3¾ inches. 72403. Length, 4 inches. 72402. Alaska. C. L. McKay.

ESKIMO SPEARS.

SEAL SPEAR.
Frobisher Bay. 10264. Captain C. F. Hall.

SEAL-SPEAR.
Four conjoined pieces; lance-point, bone, rigidly fastened into the recessed bulb-shaped end of shank; shank, walrus bone, chamfered at rear extremity and lashed to the handle with seal sinew; handle, wood; butt, recessed; small bone butt-piece inserted in recess and lashed with seal-skin; lance-strap, seal-skin. Length, 65 inches. King William Island. 10272. Captain C. F. Hall.

SEAL-SPEAR.
Pole, wood; bone spear or lance lashed to butt with a seal thong; ivory grip, carved in imitation of head of seal, lashed to central part of pole; tip of pole served with strips of baleen; head piece, ivory, recessed for ivory shank, lashed to a pole with a seal thong; harpoon wanting. Length, 9 feet 5 inches. Sledge Island, Alaska. 45418. E. W. Nelson. Harpoon and lance or spear combined.

WALRUS-SPEAR.
Detachable head, bone, tipped with slate, lashed with rawhide to a light wooden handle. Total length, 24½ inches. Alaska. 72481. C. L. McKay.

HARPOONS AND FLOAT-LINES.

HARPOON-HEAD.
Bone, with walrus-hide float-line. Eskimos, Port Foulke, Greenland. 565. Dr. I. I. Hayes.

HARPOON AND LINE.
Iron harpoon-head, with float-line made from walrus skin. Eskimos, Smith Sound. 14255. Captain C. F. Hall.

HARPOON AND FLOAT-LINE.
Line, walrus hide; head, bone, tipped with brass, fastened to line by means of a small laniard and an ivory toggle. Used by natives in capturing the beluga. Length of line, 68 feet. Alaska. 72397. C. L. McKay.

Float-Lines and Floats.

Harpoon with Bladder-Float.
Pole, wood, painted black, striped with dull red; tip served with seal-sinew and recessed for harpoon-head. Harpoon, bone, two barbed notches, attached to line with seal thong. Line probably seal sinew, stopped to pole; float bladder of seal, old, lashed to pole with seal sinew; ivory plug, ornamented, inserted in neck of bladder to be used when the bladder is inflated; finger-rest, horn. Length, 14 feet. Kodiak, Alaska. 11362. Vincent Colyer.

Harpoon-Head and Float-Line.
Detachable harpoon-head, bone, tipped with slate. Line for bending on buoy, raw-hide. Length of head, 4¾ inches. Ooglaamie, 1882. 56562. Lieutenant P. H. Ray, U. S. A.

Harpoon-Head and Float-Line.
Head, walrus-ivory, iron tip riveted with native copper. Line, walrus hide; bight caught in rigid eye of harpoon and seized with strips of baleen. Length of head, 4⅜ inches; length of line, 107 feet. Point Chaplin, Siberia. 49151. E. W. Nelson.

Seal-Skin Buoy.
Stuffed. Ornamented with ivory pendants and feathers. Flippers attached. North Greenland. 19515. G. Y. Nickerson.

Float-Line.
Line made of walrus hide. Used in capture of walrus and whales, for attaching buoys. Sledge Island, Alaska, 1880. 45403. E. W. Nelson.

Float-Line.
A line made of seal-skin, used by the natives when capturing the beluga, for bending on buoys. Cape Darby, Alaska. 48106. E. W. Nelson.

Two Seal-Skin Buoys.
Skin of a small seal turned inside out. The apertures of head and feet are tied up or hermetically fastened by means of small bone studs, with the exception of one of the fore-legs, which is used for inflation, the hole being stopped by a wooden plug. A grommet, through which the buoy-line is rove, is seized to the neck. 24 by 15 inches. 72399. 26 by 16 inches. 72400. Alaska. C. L. McKay.

SEAL DECOYS.

Three-Clawed Seal-Scratcher.
Handle and prongs wood, tipped with the claws of a seal; claws seized tightly with seal sinew, and lashed to an ivory peg,

THREE-CLAWED SEAL-SCRATCHER—Continued.

rigidly fastened in the palm. Length, 10¾ inches. Ooglaamie, 1882. 56555. Used by natives by scratching upon the ice or snow to attract the attention of seals. Lieutenant P. H. Ray, U. S. A.

FOUR-CLAWED SEAL-SCRATCHER.

Handle and prongs wood, tipped with the claws of seal; claws served with seal sinew, and lashed to a rigid ivory peg in palm; becket of seal-skin rove through a hole in the handle and knotted. Length, 8⅞ inches. Ooglaamie. 56557. Used by natives by scratching upon the ice or snow to attract seals. Lieutenant P. H. Ray, U. S. A.

REMOVING ICE AND SNOW WHEN SEAL HUNTING.

LARGE ICE-DIPPER.

Handle wood, partially painted brick-dust red; dipper made of bone, steamed and bent into almost a perfect circle (3¾ inches by 3⅞ inches at bottom, 1 inch deep), with a lip. The bottom is interlaced with seal-skin thongs, forming a strainer. The dipper is lashed to the pole with seal sinew. New. Length, 38 inches. Alaska. 36024. E. W. Nelson. Used by natives when seal hunting for removing loose ice from seal holes.

SMALL ICE-DIPPER.

Similar to 36024. Reticulated bottom, strips of baleen; handle wood, one-half inch in diameter. Old. Length, 21½ inches. Diomede Island. 63605. E. W. Nelson.

ESKIMO ICE-BRUSH.

Handle, wood; flaring bone butt-piece, inserted in recessed handle and wrapped with strips of seal-skin. Brush consists of a narrow strip of baleen, horn colored, with fringe attached, and seized to the handle with seal-skin thongs. Length, 30 inches. King's Island, Alaska. 63606. E. W. Nelson. Used by natives for brushing away snow when seal hunting, and also for brushing snow and ice from their garments.

PROBING FOR SEALS.

SNOW-PROBE.

A slender rod of bone, with a large knob and a small ferule apparently of moose-horn; ferule fastened with a small ivory peg. Length, 33 inches. Northeastern coast. 10274. Captain C. F. Hall. Used by the Eskimos in probing the air-holes in ice and under the snow to detect the presence of seals.

THROWING THE HARPOON.

THROWING-STICK.

Wood, grooved on one side; shoulder of ivory, against which the butt of the harpoon-shaft rests, rigidly fastened at the rear end of the groove. Two ivory pegs are permanently fastened on one side, at the rear end, to strengthen the grip. Used by natives for hurling the harpoon in the capture of the beluga. Length, 18 inches. Alaska. 72398. C. L. McKay.

ICE-CREEPERS.

ICE-CREEPERS.

Walrus ivory, with laniards made of seal-skin. One pair. Length, 3½ inches. Eskimo, Plover Bay, Siberia. 46260. W. M. Noyes.

WATCHING FOR SEAL.

SEAL-HUNTER'S STOOL.

Wood, heart-shaped; triangular hole cut near the center, with chamfered edge on lower sides; three small wooden pegs inserted as legs. Size, 12¾ by 8 inches; height, 5⅜ inches. Anderson River, Arctic coast. 3978. R. Macfarlane. A roughly constructed but durable utensil, used by Eskimos to stand upon while watching for seals in winter.

LINE-HOLDERS.

ESKIMO LINE-HOLDER.

A wooden rack, painted white, used by natives when beluga-fishing for carrying the line, buoy, &c. When in use it is placed on the deck of the kyak in front of the hunter. Size, 43 by 14½ inches. Alaska. 72404. C. L. McKay.

www.ingramcontent.com/pod-product-compliance
Lightning Source LLC
Chambersburg PA
CBHW020137170426
43199CB00010B/774